TRIUMPH TR7 & TR8
1975 – 1981
OWNER'S AND BUYER'S GUIDE

by James L. Taylor

1993

Published and distributed by:
YESTERYEAR BOOKS
60 Woodville Road
London NW11 9TN
(081-455 6992)

ISBN 1 873078 12 9

CONTENTS

Also published by Yesteryear Books:

Advertising Triumph (Sports Cars) 1947–1981

Also in this series:

Rover SD1 Owner's & Buyer's Guide
Range Rover Owner's & Buyer's Guide

INTRODUCTION

The TR7 and TR8 have always been controversial vehicles. From the moment the TR7 was announced in 1975, it was decried in some quarters as unworthy of the TR name: critics hated its styling, complained that it was available only in fixed-head form, and saw its carburetted four-cylinder engine as a regression from the fuel-injected six-cylinder fitted in the TR6 which it replaced. Quality control problems meant that early examples of the TR7 gave their owners a great deal of trouble. And yet the TR7 went on to outsell every single one of its predecessors in the TR sports car range.

As for the TR8, this final TR rapidly became something of a legend, even though it was introduced two years behind schedule, lasted less than two years in production, and was sold in disappointingly small numbers – and that only in the USA. That British Leyland decided to wash their hands of both this car and the TR7 in 1981 is now seen as one of the company's more regrettable mistakes – quite a contrast to earlier times, when the very introduction of the TR7 was widely viewed as a mistake!

For a long time, TR7s were not viewed as desirable cars, and many owners saw no harm in modifying and uprating them to suit their own preferences. One result of this is that relatively few cars – particularly from the early years of TR7 production – survive today in unmodified condition. When buying a vehicle, it can be helpful to know what is "standard" and what is not, for the simple reason that the factory's Workshop Manual and Parts Book will not provide assistance on non-standard items. Far better to recognise from the outset how much of a problem this is likely to be than to find out the hard way when the vehicle breaks down! One purpose of this book, therefore, is to establish "standard" specifications, to help both those who aim to keep a vehicle in original condition and those who need to establish exactly what has been modified on a vehicle.

A second purpose of this book is to point out the pitfalls associated with buying a TR7 or TR8. There are many: the poor standard of quality control on early vehicles has already been mentioned, and is the cause of several problems for today's buyers. And, of course, age has taken its toll of even the best cars. The earliest TR7s are now 17 years old, while the very last cars were built 11 years ago. Over the years, weaknesses have shown up, and a forewarned buyer is a forearmed one!

The book's format is chronological. It traces the development of the TR7 and TR8 ranges from 1975 to 1981, discussing in detail each stage in the vehicles' development, and providing specification change data and other useful information. After each of the descriptive sections comes a detailed guide on the strengths and weaknesses of the models in question. The advice the book contains is designed not only to assist buyers in their choice of model, but also to warn them about potentially expensive faults. However, because major items like engine or gearbox are dealt with only once (on their first introduction), it is important to read *the whole book* and not just the section relating to a model of particular interest.

For interest, the book also includes brief details of the cars' competitions career and looks at planned derivatives which did not go into production. A further section covers modifications commonly made to TR7s and TR8s. Finally, the book includes some helpful specification tables and statistical data which will be of interest to the TR7 or TR8 owner.

Although neither the author nor the publisher can accept any liability for acts carried out on the advice given in this book, both would be pleased to be informed of any errors or omissions which have remained undetected.

October 1992   James L. Taylor

THE TRIUMPH MARQUE

Triumph first established a reputation as a motor cycle manufacturer, adding cars to its product range in 1923. The car and motor cycle sides of the business were separated in 1936, however, and the car business went into receivership three years later. A solution to the company's problems was delayed by the 1939–1945 War, but in 1945 Triumph resurfaced under the ownership of Standard.

In the early 1950s, Triumph introduced the first of its TR sports roadsters. These sold extremely well in the USA and, at home, reinforced the sporting image associated with the marque since the 1930s. Standard, meanwhile, had become bogged down with a series of rather stodgy saloons and so, after 1959, all new cars from the Standard-Triumph combine were given the more evocative Triumph badge.

However, even the marque's export successes were not enough to prevent a second major financial crisis in 1961. This time, the saviour was Leyland Motors, which bought the Standard-Triumph combine as a going concern. During the 1960s, both saloons and sports cars with the Triumph name continued to sell successfully at home and abroad.

Major changes affected Triumph in the early 1970s. After the series of mergers within the British motor industry during the late 1960s which produced British Leyland, several formerly competing marques had now come together under the same ownership. Triumph, for example, was in conflict with Rover for its larger saloons, with Riley, Wolseley and Vanden Plas derivatives of Austin and Morris models for its smaller saloons, and with MG for its sports cars. Clearly, product rationalisation was called for.

So it was that the early 1970s saw the Triumph engineering department merged with that of Rover, while sports car expertise from MG also became available to the Triumph engineers. The TR7 and TR8 were both products of this rather confused period.

Nevertheless, the Triumph marque did not prosper during the 1970s. The failure of the Stag grand tourer, the problems associated with the TR7 family, and the cancellation of the cars planned for the 1980s left little of any value associated with the Triumph name. The last car to bear Triumph badging was the Acclaim, a small saloon built under licence from Honda between 1981 and 1984. Its successors wore Rover badges.

The car which the TR7 had to replace was the TR6, a rorty six-cylinder machine in the traditional British roadster mould. It is seen here wearing the optional hard top.

ORIGINS AND DEVELOPMENT

The origins and subsequent development of the Triumph TR7 and TR8 sports cars are an immensely complicated story all of their own. The cars were conceived at the end of the 1960s, when the newly-formed British Leyland was struggling to bring under corporate control the many different marques it had acquired, and that struggle caused several changes of plan for the cars themselves. They entered production during BL's bleakest period in the mid-1970s, and their planned development was so seriously disrupted by labour troubles that they never achieved their anticipated sales volumes and were therefore taken out of production early.

The British Leyland Motor Corporation was formed in 1968, and brought together under a single umbrella all three of Britain's principal sports car manufacturers – MG, Jaguar and Triumph. As all three had flourishing model ranges, BL management allowed them to carry on undisturbed for the first two years of the new corporation's existence; but once it became time to think of new models, that management quite sensibly saw an opportunity to rationalise its model ranges in the interests of greater cost-effectiveness.

Jaguar was to some extent a special case, as it was also a successful manufacturer of luxury saloons, and so the BL management decided to leave it well alone for the time being. However, there was little point in developing two new sports cars – an MG and a Triumph – to compete in the same market sector and to prolong the traditional rivalry between the two marques. Costs could be saved by developing a single model to replace both the Triumph TR6 and the MGB and, by 1970, BL management had resolved to do exactly that. It was symptomatic of the company's problems at the time that both MG and Triumph engineers were already working on completely separate designs for new models, with the result that a great deal of time and effort had already been wasted.

About one thing, however, the BL management was quite clear: most important in the specification of its new sports car would be the requirements of the North American market, which for many years had taken the majority of MGs and Triumphs produced and would therefore be the foundation on which sales of any future model would be built. Yet there was already a threat to these sales in the shape of two new rivals – the Datsun 240Z and the VW–Porsche 914 – which had appeared on the market in 1969 and had captured the imagination of American sports car buyers. BL therefore decided to start from a clean sheet of paper with a re-examination of transatlantic requirements and, in late 1970, it conducted a survey in the USA to find out what characteristics would be necessary in an ideal medium-sized sports car for that market.

The survey suggested that advanced styling, good comfort, and high equipment levels should be allied to simple, reliable, and easily-serviced mechanical components. And, of course, the new car would have to conform to all existing and planned US legislation on crash safety and exhaust emissions control. As the car already on the drawing-boards at MG was a mechanically complex mid-engined design, it was obviously not going to fit the bill, and was abandoned in favour of the simpler Triumph plans. As Triumph's engineering facilities were more modern than MG's, BL management also decided that Triumph engineers would develop the new sports car, although this would rule out neither contributions from other parts of BL nor the possibility of badge-engineered variants of the same design. So it was that the new car would be very much a Triumph, even though other parts of the BL empire did make significant contributions to its design throughout the project.

In fact, Triumph as such ceased to exist very soon after the go-ahead had been given to its plans for the new sports car. BL merged its engineering department with that of Rover during 1971, and drew up a new, integrated, Rover–Triumph product plan for the 1970s. Under this, certain key engineering features would be shared by four new models. The first of these models was Bullet – Triumph's codename for its two-seater sports car. The second was Lynx, a 2+2 long-wheelbase coupé derivative which had always been part of the Triumph plan and was aimed at the grand touring market into which the Stag had just been launched. The third car was a big new saloon which would replace the Rover P6 and Triumph "Innsbruck" ranges; and the fourth was a new medium-sized saloon which would replace the Triumph Dolomite range. The latter two were given the project names of SD1 and SD2 respectively when Rover–Triumph was re-christened the Specialist Division of BL in 1972.

The merger with Rover made available to the Bullet and Lynx projects that company's well-proven 3½-litre V8 engine, and this became an engineering cornerstone of both cars. In Lynx, it would be the only engine option; but for Bullet, it would be an alternative to the Triumph Dolomite-derived 2-litre slant-four. Gearboxes and axles would also be shared with other Rover and Triumph models, although development lead-times made clear that Bullet, as the first of the four new models scheduled for production, would initially have to use existing components from elsewhere within the BL empire.

As for styling, the threat of legislation to outlaw open cars in the USA persuaded BL to settle on a fixed-head body for Bullet. The original Triumph proposals were rejected in favour of a wedge-shaped two-seater design by Harris Mann of the Austin styling studios at Longbridge, and Lynx was planned as a derivative of this.

Development of the four-cylinder (TR7) and V8 (TR8) versions of Bullet went ahead in parallel from the beginning, even though it was clear that the TR8 would have to be introduced after the smaller-engined car because the stronger gearbox and axle it needed would not be ready in time for simultaneous launches. Prototypes of both cars were running during 1973, and work went ahead rapidly. By the end of 1974, the TR7 was ready to go into production. BL had decided that it should be built at the Triumph factory at Speke, which was ideally situated close to Liverpool with the excellent port facilities vital for a car which would depend so heavily on export sales.

When the four-cylinder TR7 was launched in 1975, work had already begun on development of other versions to be introduced later. The TR8 was just one of these. Also under development was a more powerful version of the TR7, known as the TR7 Sprint, which had the 16-valve version of the slant-four engine used in the Dolomite Sprint saloon. And before long, these two proposed models were joined by a new project to develop a drophead body for the TR7/TR8 family.

This about-face had been caused by the failure of the proposals to outlaw open cars in the USA. As the TR7's fixed-head styling was controversial, and there was no doubt that an open TR7 had enormous sales potential, BL moved rapidly in 1976–77 to develop a drophead body. In the early stages, it seems that conversion specialists Crayford Engineering were asked to put forward a design, although the final production body was developed in-house.

Meanwhile, however, the Speke factory had been suffering from constant labour relations troubles, and in the autumn of 1977 its workforce staged a four-month strike. The TR7 drophead, TR7 Sprint, TR8 and Lynx were all ready to go into production, but BL management decided to close the Speke factory and relocate TR7 production in the Midlands. The introductions of the drophead body and of the TR8 were merely postponed, but the production of the TR7 Sprint and the Lynx was cancelled altogether.

The drophead was therefore the last variant of the TR7/TR8 family to be designed, even though the complexities of BL's problems in the 1970s ensured that it was the V8-engined TR8 which was actually the last variant to go into production. Although a re-engined TR7, with the then-new O-series 2-litre four-cylinder engine, was under development for the 1982 model-year, the continuing strength of the pound against the US dollar put BL into a position where it was forced to make a painful choice: it could price the TR7 out of a market which took 80% of all cars produced, it could sell them at less than they cost to manufacture, or it could stop production altogether. It chose the latter course, and the TR7/TR8 production lines closed down on 5th October 1981. Ironically, within a month of the announcement that production was to end, the pound had slumped dramatically against the US dollar and other currencies.

This is the original styling sketch for the TR7, penned by Harris Mann of the Longbridge styling department. The TR7's detractors have claimed that the sketch was little more than a doodle and was not a serious styling proposal at all.

CONFIGURATION

The basic configuration of the TR7s and TR8s was the same: all of them had saloon-derived mechanical components in a two-seater monocoque body shell. The fact that they shared mechanical components with more run-of-the-mill vehicles anchored them firmly within the tradition of the earlier TR models, and yet the monocoque shell was a new departure for the TR range, for all earlier models had been constructed with a separate chassis and body.

Bodyshell and styling

Right from the beginning, the most controversial feature of the cars was undoubtedly their shape. The later drophead bodies attracted almost universal acclaim, but the fixed-head body with which the TR7 was announced in 1975 gained the car few admirers.

There was little to object to at the front end, and most early press reviews of the TR7 approved of the sloping-nose treatment which gave the car a dramatic frontal appearance. Aft of the bonnet and steeply-raked screen, though, was where the styling became questionable. The very short wheelbase, dictated by the need for handling agility and by the fact that the TR7 was an uncompromising two-seater, gave the car a stubby appearance from the side, which was not helped by the short, squared-off, boot. The near-vertical rear window also did nothing to help the car's looks, and the rear lamp cluster design was simply clumsy. Along the flanks, an upward-sweeping crease emphasised the wedge-shape while drawing attention away from the high waistline, but it was never much liked and plans to delete it were under consideration when TR7 and TR8 production stopped in 1981.

Many of the car's features had come about because of the need to meet US safety regulations. Thus, for example, the forward-hinged bonnet was fitted with a safety interlock to prevent it from slicing through the windscreen in a frontal collision. Similarly, the head-lamps were concealed in pods which were raised electrically from the streamlined front, because this was the only way Triumph had been able to get lamps at the height required by the US regulations without spoiling the bonnet shape.

Under the skin, it was US crash-safety regulations which had also dictated much of the bodyshell's structural design. In fact, Triumph had produced a shell which was remarkably light in view of the strength engineered into it, but it was true that it appeared to the untutored eye to be rather wasteful of space. Although the car had only two seats and a fairly small boot, it was both longer and wider than the Dolomite saloons current when it was introduced, and these were full four/five-seaters.

In fact, a good few inches of length were sacrificed to the space wasted ahead of the radiator in the sloping nose, and to the massive bumpers. These consisted of a resilient black polyurethane outer skin over a steel armature, and had been designed to meet the anticipated US legislation under which bumpers would have to withstand a 5mph impact and protect the rest of the car from damage. In order to meet the 5mph criterion at the rear, the bumpers wore huge and ungainly integral overriders, which once again did nothing for the car's appearance. Ironically, the 5mph impact regulation never did become law, and by the time the TR7 was introduced, the US legislators had settled on a 2½mph impact rule.

Interior

The great torsional strength of the TR7's bodyshell gave it a feeling of solidity which was more like that of a Mercedes-Benz sports car than that of the "traditional" TRs. The interior layout reinforced this impression, with its large moulded dash panel, high-waisted doors, and discreet but recognisable crash-padding. Yet there was more room in the TR7 than its exterior dimensions suggested. Its seats were well-shaped and extremely comfortable, with reclining backrests and adjustable head restraints. There was a large drop-down glove-box on the passenger side of the facia, although storage space was otherwise limited, with map pockets in the body sides, a small rear parcels shelf, a lidded box on the centre console, and recesses on either side of the dash top.

It was quite obvious that the dashboard had been designed for easy installation of either left-hand or right-hand steering, but the instrument panel which fronted it was commendably neat. Quite unlike that of previous TRs, this was now injection-moulded to meet the crash-safety regulations which had brought about the demise of the wood finish associated with Triumph sports cars since the TR4A of the mid-1960s. Even the steering wheel had a soft, collapsible, rim and – on early US models – its centre was covered by a huge and ugly crash-pad.

Creature comforts were well catered for in the heating and ventilating department. The TR7 and TR8 used a "corporate" BL heater (it had first been seen in the 18/22 – later renamed Princess – saloons, and was also used in the Rover SD1), which was both effective and simple to regulate. Within the same heater box could be fitted an optional air-conditioning unit, which used the same dashboard vents as the standard heater. When this was fitted, an extra heat exchanger radiator was installed ahead of the engine coolant radiator, and cooling was assisted by twin electric fans mounted right in the nose of the car.

Mechanical elements

A good proportion of the TR7's and TR8's mechanical components were borrowed or adapted from those in Triumph's Dolomite saloons. In the case of the suspension, the Dolomite's layout provided the TR7 with much better ride and handling qualities than any previous TR sports car had enjoyed. At the front were MacPherson struts, with coil springs and an anti-roll bar; both engine and front suspension were carried on a small sub-frame which bolted to the monocoque. At the rear was a coil-sprung live axle with two longitudinal trailing arms and two semi-trailing arms, the whole suspension being bolted directly to the bodyshell. The TR7 installation improved on the Dolomite, however, by providing a full eight inches of wheel travel, which permitted a relatively softly-sprung ride with the minimum risk that the suspension would bottom out over potholes or corrugations. (This long wheel-travel was a trademark of Spen King, then British Leyland's technical chief, who had earlier applied the principle successfully in his Rover 2000 and Range Rover designs).

The short wheelbase gave taut handling, ably assisted by a sharply responsive rack-and-pinion steering system. Brakes were again similar to the Dolomite layout, with identical rear drums but larger-diameter front discs. Power assistance came from a direct-acting servo, and there were split-circuit hydraulics, with a pressure limiting valve in the rear circuit to maintain front-to-rear balance. Clutch, four-speed gearbox, and rear axle ratio were all Dolomite-derived (although the gearbox had originally been designed for the unloved Morris Marina!), and in the drive-line only the shorter propshaft and the absence of the saloon's optional overdrive were differences.

It was to the Dolomite again that Triumph had turned for the TR7's engine, although the version installed in the sports car was a hybrid of existing units. On to the 1998cc (122 cu.in.) cast-iron block of the twin-cam, 16-valve, Dolomite Sprint was bolted the single-cam, 8-valve, aluminium alloy cylinder head of the "cooking" 1854cc unit. The 16-valve engine itself had not been used because it had not yet been satisfactorily redeveloped to meet US exhaust emissions control regulations, but the extra capacity of its big-bore block gave both additional power and torque over the 1854cc engine. The TR7 engine also had the advantage of a transistorised, contactless, ignition system.

Although the TR7 engine as such was therefore new, the elements which went into its make-up were not. The original design of the engine could be traced back as far as 1963, when what became the Dolomite engine had been sketched up as part of a family of new engines. The larger ones of these were to be V8s (although the only one to see production was that in the Stag), while the smaller ones were four-cylinders, which were canted over at an angle of 45°, and were in essence half of the V8. This slant of course allowed the relatively tall engine to clear a low bonnet-line, a feature which helped to make it an ideal choice for the TR7. The slant-four had first been built in 1796cc form for Saab in the late 1960s, but the first Triumph car to have it was the Dolomite in 1972.

Even though Triumph had its own V8 engine, it was not this which would be used for the TR8. For reasons which have already been explained, the choice of the TR8 engine fell upon the Rover 3,528cc V8, which was actually a General Motors (Buick) design to which Rover had bought the manufacturing and development rights in 1964. The TR8 version of the engine, and the gearbox and axle which went with it, came from the Rover SD1 saloon which was developed in parallel with the TR7 and TR8 cars. The TR8's five-speed gearbox and rear axle (albeit with different gearing) were in fact fitted to TR7s for some time before TR8s became available, and replaced the older and less robust Dolomite-derived items.

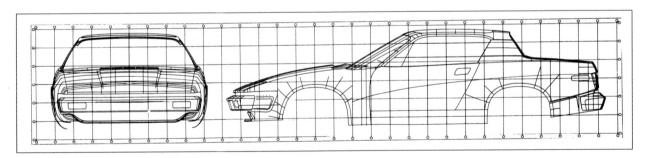

Computer-aided design was coming of age during the TR7's lifetime. These side and front elevations of the bodyshell were produced on an automatic drafting machine at the Leyland Cars Engineering Data Centre in the mid-1970s.

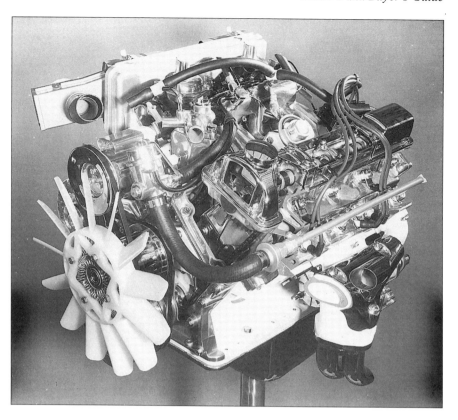

A cutaway display TR7 engine. In this picture, the engine's "slant" towards the camera is quite clear. The basic design had of course been drawn up in tandem with that of the V8 used in the Stag, and the slant-four cylinder block was effectively the left-hand half of a V8 block. Note how height is minimised by the position of the inlet manifold and carburettors on the right-hand-side of the engine (i.e. further from the camera in this picture).

What looks like a bulky engine when out of the car is almost lost in the TR7's wide engine bay which was, of course, also designed to take the Rover V8.

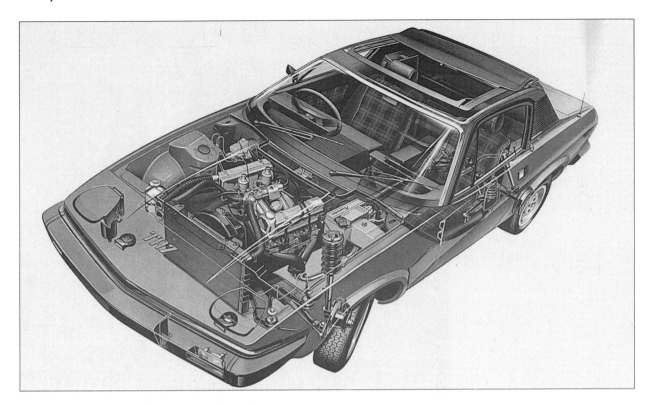

This cutaway drawing of the TR7 appeared in early sales catalogues. Note the layout of the MacPherson strut front suspension, and the amount of unused space in the nose ahead of the radiator.

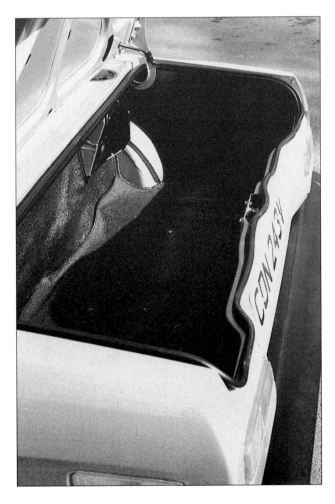

The TR7's boot was much larger than it looked. BL claimed its capacity was 10.3 cubic feet.

When the car's lights were off, the headlamps were concealed in the sloping nose. When the lights were switched on, electric motors raised the headlamps to their operational position.

This rear view of a late TR7 drophead shows the American Federal-specification bumpers with their integral overriders. These were not fitted on cars for other markets.
(Picture by courtesy of Dave Destler at British Car magazine.)

THE SPEKE-BUILT TR7, 1975-1978

In view of the importance of North American sales to the TR7, it was only to be expected that BL would launch the car in the USA before making it available elsewhere. The pilot-build models were completed at the end of 1974, and a number were shipped across the Atlantic to Florida, where they were shown to the Press at Boca Raton on 15th January 1975. Sadly, labour troubles at the Speke factory had already begun to take effect, and some of the cars had been hastily finished off with sub-standard components. Showroom sales in the USA began on 2nd April 1975.

As for the British Press, BL showed a few pre-production cars to selected journalists just before the American launch, in order to ensure that parallel press coverage in the UK would whet the buying public's appetite for when production examples became available later. The original plan was in fact to launch the European versions of the TR7 at the Paris Motor Show in autumn 1975, but demand from North America was so high that the launch had to be postponed. The first postponement was to the Geneva Show in March 1976, but there was then a second postponement, so that the eventual launch was not until 19th May 1976.

All these first TR7s were of course fixed-head cars: the drophead model would not become available until 1979. For all markets, they had decal-type badges front and rear. That on the nose simply read "TR7", while at the rear the car's full name was spelled out on the lower edge of the boot lid. There was also a small enamelled BL logo on the front wings, just behind the wheel arches. All body glass was tinted, and all windscreens were of laminated glass. The first cars had silver-finished wheels with ventilating slots and black centre finishers, and the wheels were shod with 175/70 x 13 low-profile radial tyres. A black-finished driver's door mirror was also standard.

There were only minor interior differences between cars for the US market and those for other markets. All cars had seats upholstered in "broadcord" nylon cloth (a sort of corduroy), but only US-specification cars had a steering wheel which incorporated a large crash pad in its centre: those for other markets had a three-spoke type. On US-market cars, there was a service interval indicator lamp for the charcoal canisters which formed part of the exhaust emissions control system, but this lamp was not fitted for other markets. Air conditioning was also available only on US-market cars, and that as an extra-cost option.

Body differences between US-market and other cars were rather more marked. The running lights in the front and rear wings of US models were not fitted for other markets, where their apertures were filled by black plastic blanking plates. Outside the USA, the massive rear overriders were not fitted, although the

bumpers were otherwise similar. Lastly, only the US models had a special narrow fuel filler neck to suit the pump dispenser nozzles used for lead-free fuel in that market.

The most significant differences between the US-specification cars and those sold elsewhere related to the engine. For the USA, there were actually two states of engine tune: a "49-State" or "Federal" tune and a lower "Californian" tune which met the stricter exhaust emissions regulations which applied in that State.

Outside the USA, the 2-litre TR7 engine had a compression ratio of 9.25:1, twin SU HS6 carburettors, and developed 105bhp at 5,500rpm. In "49-State" form, it developed just 90bhp at 5,000rpm as a result of modifications made to meet emissions-control regulations. Among the changes were a lower compression ratio of 8:1 and twin Zenith-Stromberg carburettors instead of the twin SUs. In "Californian" form, it developed no more than 76bhp and had just one Zenith-Stromberg carburettor. Not suprisingly, the cars sold outside the USA had by far the best performance, both in terms of acceleration and maximum speed. However, the higher crankshaft speeds of the engines not fitted with emissions control gear did enhance rather than detract from an unfortunate tendency to fussiness at high crankshaft speeds.

As a result of the exhaust emissions control equipment fitted to its engine, the US-model TR7's performance was rather disappointing, and many potential customers appeared to consider that it was not sufficiently sporting. The US dealers responded to this very quickly by asking BL for an exhaust system with a crisper note, so that the car would at least *sound* like a sports car! They got what they asked for, although the original, quieter, system remained standard on TR7s for other markets.

Outside the USA, it was quite clear that the TR7's handling and roadholding were too good for the performance available, with the result that the cars could feel underpowered and rather characterless. However, this tendency was not increased by the lower-powered engines of the US models for the simple reason that their handling was not quite as good: BL had fitted softer suspension bushes to suit the general American preference for a boulevard ride. Otherwise, all aspects of the drivetrain and running gear remained the same for all markets.

For the 1977 model-year, however, two new transmissions were introduced. One was a Borg Warner type 65 three-speed automatic, which brought with it a taller 3.27:1 final drive, while the other was the brand-new five-speed "77mm" manual gearbox which had first been seen in the Rover SD1 3500 saloon in June 1976. The new five-speed gearbox came with a narrow-track version of the Rover's back axle and a special 3.9:1 final drive, which aided acceleration while the overdrive fifth gear maintained relaxed cruising and reasonable fuel economy figures. The Rover axle, which Triumph literature always described as a "heavy-duty" axle to distinguish it from the "medium-duty" Dolomite type, also came with larger-diameter brake drums and wider 185/70 HR 13 tyres, which were fitted to the front wheels as well.

As this stage, however, the five-speed gearbox had only just entered production and build volumes were not enough to cope with the demand from both Rover SD1 and Triumph TR7 assembly lines. Following its usual policy of catering primarily for the US market, BL therefore standardised the new gearbox and axle on TR7s sent there, and introduced them in the UK as an extra-cost option in September 1976. In fact, relatively few five-speed cars were built for the home market, and it looks very much as if BL's intention had always been to build just enough cars to get the gearbox and axle homologated for use in the works rally TR7s (existing rules demanded that gearboxes used in competition cars had to be "production" items, and "production" was defined as a minimum quantity built for sale in the UK market). In December 1976, therefore, home-market TR7s reverted to the four-speed Dolomite gearbox and medium-duty axle, and the five-speed option was deleted. Over in the USA, of course, the five-speed transmission remained standard.

Changing exhaust emissions control regulations in the USA meant that the 1977-model TR7s for that market had to have catalytic converters in their exhausts instead of the charcoal canister filters fitted to earlier cars. But this and the complications of the alternative transmissions were enough for BL to cope

with at this stage. The only other change for 1977-model TR7s was that the small BL logos had now been deleted from the front wing bottoms. Exactly why has never been clear, but it may well be that the corporation's poor public image had something to do with it; perhaps the marketing men thought it better not to remind the customers that the TR7 was actually a British Leyland product!

The next series of revisions came in March 1977. Cars built after this date could be recognised by their silver wheel-centre trims, and by new upholstery material with vinyl side panels flanking plaid cloth wearing surfaces. A folding fabric sunroof also became available as a factory-fitted option. In addition, the TR7's ride height was lowered by one inch at the rear through the use of shorter coil springs with different spring rates. The resultant effect on the car's appearance was surprisingly successful, and the lowered rear end lessened the stubby, high-tailed, look and made the TR7 look both longer and more streamlined.

During this period, BL also marketed a limited-edition model known as the Jubilee, capitalising on the fact that 1977 was the Queen's Silver Jubilee Year. The main distinguishing features of this model were the twin wide stripes which ran along the body sides, and the "TR" decals on the doors just ahead of the handles.

Production of the five-speed gearbox had reached high enough levels by the summer for it to be reintroduced as an option for markets outside the USA in October 1977. In practice, however, the Speke strike which started that month ensured that there were very few 1978-model cars of any sort, with the result that the five-speed gearbox remained rare outside the USA until TR7 production was transferred to Canley in the summer of 1978.

The first TR7s were built for the American market. In this view of an early Federal-specification example, the side marker lights required in the USA are clearly visible.

Leyland Cars went through a period in the mid-1970s when it issued press pictures in which models attracted almost as much attention as the cars! This picture shows an early UK-market car. The cutouts in the wings for the Federal specification marker lights were filled by ribbed black plastic blanking plates.

This rear view of the same car shows the controversial coupé roof with its vertical rear window. UK-market cars did not have the integral overriders which Federal models were obliged to have on their rear bumpers.

The passenger compartment of an early right-hand-drive TR7. Note the clear instrumentation and the broadcord seat facings.

The automatic transmission version of the TR7 remained relatively rare, and could only ever be distinguished by looking inside.

This type of bootlid badging was only ever used on cars built at Speke.

The original Federal TR7 steering wheel had a massive crash pad in its centre. Also visible here is the Speke style of lettering on the glove-box lid, with lower-case letters.

BUYING A SPEKE-BUILT TR7

There can be no doubt that the Speke-built TR7s are the least desirable of all the TR7/TR8 family. Build quality was almost always poor, and only too often appalling. One result is that the cars are now likely to be suffering from minor but annoying faults which previous owners have considered too costly or time-consuming to rectify. It also means that some cars will have been refurbished: remember that it is very easy to respray a car and to replace the distinctive "TR7" decal on the nose with the wreath badge which was associated with the later and better-built cars. If in doubt about the vintage of an apparently later car, check the commission number against the list given in the section on Vehicle Identification at the end of this book.

In Great Britain, the early TR7s will be those with P-, R-, and S-suffix registrations. A very small number of US-specification left-hand-drive cars was put on the road by BL with N-suffix registrations before the European launch, but it is unlikely that any of these still survive; poor build quality and left-hand-drive will have conspired to bring about their early demise.

When buying an early TR7, the most important thing to consider is overall condition: the specification differences between models were not great enough to make any one model significantly more desirable than another regardless of condition.

Nevertheless, it is worth examining those specification differences which can influence a purchase. As far as transmissions are concerned, there are four-speed, five-speed, and automatic models: the automatics are fairly rare and are not highly regarded because they lack sporting character, while the five-speed manual box is very much preferable to the rather notchy four-speed item. It has a much nicer change quality and ratios which are better suited than those of the four-speed to the engine's power and torque characteristics, it offers better fuel economy, and it gives the TR7 a higher maximum speed of around 115mph. Unfortunately, as the previous section makes clear, five-speed boxes are rare on early cars except in the US market. All four-speed and automatic cars were delivered with 175x13 tyres, but some owners have fitted the 185/70HR13 size associated with the five-speed cars. This need not cause concern, because both types of tyre fit on the same rims.

The other main specification difference relates to the upholstery material, and there is little doubt that the plaid type used after March 1977 both looks nicer and wears better than the early broadcord type, on which excessive wear can create bald patches in the pile. Fabric sunroofs were available from the same date, but may of course have been fitted retrospectively to earlier cars.

Body rust affects these early TR7s quite seriously, and can be easily disguised by a respray. As the body tooling for the TR7 was scrapped in the mid-1980s, supplies of factory-made replacement panels are now drying up. Some will become unavailable, and others will remain available only if remanufactured by after-market specialists. Some of these specialists have now produced patch panels to repair the most common problem areas of a TR7. However, it is important to remember that the limited numbers in which all these remanufactured parts are made means that prices will never be low, with the result that any TR7 offered for sale which needs major body restoration is probably best avoided on economic grounds.

Rust begins at the bottoms of the front and rear wings just behind the wheelarches, and in the horizontal welded seam between the rear wheelarch and the bumper. Leaking boot seals can also lead to rust in the boot floor, and all folded-over body seams are vulnerable. Paint peels off the headlamp shells, which are made of aluminium and on early cars were not coated with the appropriate etching primer before being painted. The edges of the under-bumper radiator air intake also rust badly after stone chips remove the paint and expose bare metal. As the sloping bonnet makes it impossible to see the front of the car from the driving seat, parking tends to be a matter of practice and many TR7s will have minor bumper damage. Some buyers may be prepared to tolerate small blemishes: others should remember that replacement bumpers are expensive!

It is worth emphasising that the TR7 is a strict two-seater, and that interior space is very limited indeed: even a briefcase will have to go in the boot if both seats are occupied, and there is definitely no room for child-seats, carry-cots, or the like in the back. Assembly faults in the interior were usually confined to poorly fitting parts in the dashboard, but even more common was a badly-applied TR7 sticker on the glove-box handle; these stickers tend to peel off.

As for the boot, few owners have ever complained that it is too small; and BL claimed that it would hold not only a set of golf clubs but the trolley as well. Unfortunately, golf trolleys and other bulky items tend to have sharp edges, and the interior of the TR7's boot is not lined. In consequence, minor damage sometimes shows through on the outer skin of the rear wing panels.

The cars also have their fair share of engine problems. The TR7 engine is prone to blowing head gaskets, and both water loss and general rough running can warn that a gasket has blown. The TR7's tendency to blow gaskets has led to some engines being "cooked" and, as with the related Stag V8 engine, it is possible to disguise a warped cylinder head by using a thick layer of gasket cement. Such cheap "repairs" always fail sooner or later, usually leading to a repair bill larger than would have been incurred if the job had been done properly in the first place. Worth remembering is that all standard production varieties of the 2-litre engine left the TR7 underpowered for its handling abilities, with the result that many have been thrashed quite mercilessly.

Particular points to be wary of are noisy timing chains (possibly caused by low oil pressure or by a stretched chain which is just about to jump its timing-wheel sprockets), and overheating. This latter might point to imminent or poorly-repaired head gasket problems, although the cause might equally be nothing worse than a blocked radiator core. When coupled with low oil level or oil pressure problems, overheating can quickly lead to main bearing failure.

Carburation problems are relatively common on TR7s, whether equipped with twin SUs, twin Zenith-Strombergs for the US "Federal" market, or a single Zenith-Stromberg for California. The SUs are generally reliable; the Zenith-Strombergs rather more temperamental and awkward to tune. One of the most common problems on the twin-carburettor engines is rough idling, which can often be cured by simply topping up the oil in the carburettor dashpots; another common cause is wear in the rubber carburettor mountings. An abnormally high idling speed which cannot be cured by straightforward carburettor adjustment may well be caused by sticking over-run valves in the carburettor butterflies. However, these are the simple solutions to carburettor problems: intractable tuning difficulties can often be caused by very small induction leaks, and these can be extremely difficult to trace.

It is an unfortunate fact that the early emissions-controlled engines often feel under par even when running perfectly, and there is nothing which can be done about this: the single-carburettor Californian cars are the worst offenders here. Rough running and poor idling on the Federal models may often be caused by nothing worse than incorrect balancing of the twin carburettors, but that in itself could also mask rough running attributable to a blown head gasket!

Faults also surface in the rest of the power train. The four-speed gearboxes were notchy when new and do not improve with age; synchromesh on the lower ratios was always marginal and may have ceased to work altogether on a high-mileage car. Five-speed types often suffer from obstructive synchromesh when cold, especially between first and second gears, and an approved remedy for this is to replace the ordinary gearbox oil with automatic transmission fluid. Five-speed gearboxes can also suffer from worn layshaft bearings, revealed by a chattering noise at idle. Further back down the driveline, axles are often noisy on the over-run, but this fault was present when the cars were new and need not herald a major repair bill.

Ride and handling were the TR7's strongest suits, and any deficiencies in these areas need investigation. Worn suspension bushes are the most common problem. Steering wander, shimmy, rapid tyre wear, or rattles are all indicative of worn bushes in the front suspension. Dubious cornering behaviour and clonks are the tell-tale signs of similar wear at the rear. All early TR7s were recalled for attention to the rear suspension arm mounting brackets, and it is advisable to check that all is well here. Brakes are mostly trouble-free, although some drivers find them marginal and it is true that the front discs can wear quite badly with heavy use.

Electrical problems are common, and the TR7 was one of those models which caused frustrated American buyers of BL cars in the 1970s to refer to Joseph Lucas (founder of the electrical components company) as the Father of Darkness. Faults can occur in the transistorised ignition and in the mechanism for raising and lowering the headlamps. In the latter case, the problem is often not primarily electrical: water and dirt can find their way inside, causing the mechanical linkages to jam, relays to blow or, in the worst cases, the motors themselves to burn out. Finally, lackadaisical work by some BL garages did nothing in the early days to prolong the useful life of the first TR7s.

If this sounds like a catalogue of disasters, it is nevertheless true to point out that not all early TR7s were badly made: a few good ones did slip through the net and some were kept in good condition by fastidious owners. Such cars are very much exceptions to the rule, however. Always remember that a rough early TR7 is likely to be a constant source of trouble and disappointment, and is unlikely ever to repay time and money invested in putting it right. The early TR7 is not yet well-regarded, and it will probably be a long time before the car attains the classic status accorded to earlier TR models and prices go up correspondingly.

THE 1977 PRODUCTION STOPPAGE

Ever since the first TR7s had come off the production line at the end of 1974, the Speke factory had been plagued by industrial relations troubles which had made meeting production targets difficult and had done nothing to promote the sort of atmosphere in which build quality problems could be tackled. Things came to a head in October 1977, when a strike completely halted production. TR7s began to come off the lines again in February/March 1978 after a stoppage of sixteen weeks, but labour relations were still not healthy, and BL (in the person of the then Chairman, Sir Michael Edwardes) decided to cut its losses and close the Speke factory altogether, transferring TR7 assembly to the older Triumph factory at Canley, near Coventry. The last Speke-built TR7 was made in May 1978, and there was then a second production hiatus until the Canley lines became operational in October 1978. Between the beginning of the Speke strike in October 1977 and the resumption of production at Canley a year later, no fewer than eight months had been lost out of the 1978 model-year. There were, in consequence, very few 1978 TR7s.

Worse than the loss of production, however, was perhaps the fact that this disruption on the lines caused BL to cancel two new TR7 derivatives, and to postpone the introduction of two more. Completely cancelled were the Lynx and the TR7 Sprint, while the need to build up production of the base model again meant that there was no room on the assembly lines for two even more important models – the convertible version and the V8-powered Triumph TR8 – and so their introduction was held over. Prototype or pre-production examples of all four models were in existence at Speke before the closure.

The *Lynx*, of course, was the companion model to the TR7 which had been planned right from the beginning. In Triumph's model range, it would have replaced the Stag, which had gone out of production in mid-1977 but was still available in BL showrooms when the Speke strike began. In essence, the Lynx was a 2+2 luxury grand touring coupé, but with a fixed roof instead of the interchangeable hard and soft tops of the Stag. Its styling had been agreed as early as 1974, and was the same as that of the TR7 as far back as the

windscreen pillars. Behind that, there was a longer wheelbase to accommodate the extra pair of seats, and the TR7's controversial side scallop was deleted in favour of a neater indentation feature. The roof swept into a pleasing fastback shape. Mock-ups of the interior indicated that trim would be plusher than in the TR7, and the plan was to offer the car with Rover's 3½-litre V8 engine and either the five-speed gearbox or three-speed automatic. Around six Lynx prototypes are thought to have been built. Two have been preserved by the British Motor Industry Heritage Trust, but the others were probably all broken up.

The *TR7 Sprint* was simply a TR7 fitted with the 16-valve engine from the Dolomite Sprint. In saloon form, this engine developed 127bhp at 5700rpm, and the extra 22bhp over the standard (European) TR7 engine would certainly have given the sports car the sort of performance which its looks suggested. Whether the TR7 Sprint would ever have appeared on the American market is not clear: one reason why the TR7 did not have the 16-valve engine from the beginning was that Triumph were having difficulties in getting it to conform to North American exhaust emission standards. At one stage, it was rumoured in the British Press that the TR7 Sprint would be launched instead of a V8 model in Europe. Around 25 genuine TR7 Sprints are thought to have been made, and most were probably sold off through the trade by British Leyland. Not all 16-valve TR7s belong to this rare breed, however, for some specialists have offered "Sprint" conversions for ordinary TR7s.

The *TR7 drophead* did go into production eventually, of course, as did the *TR8*. But it is interesting to speculate whether the TR7/TR8 range would have been killed off so abruptly in 1981 if the planned range expansion had gone ahead and had garnered the increased sales which British Leyland hoped for in the mid-1970s. And yet, in at least one way, the Speke strike was good for the TR7 and TR8, because it forced BL to tackle the problem of poor quality control. The result was a much-improved range of cars, which showed the strengths of the TR7s and TR8s to the best advantage.

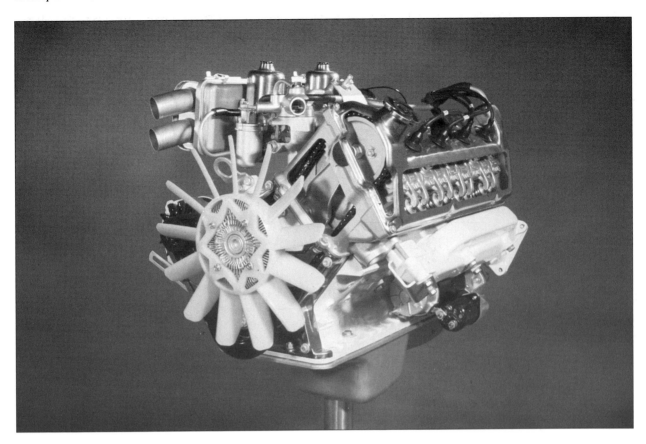

One of the cars which was almost ready for production when the Speke workforce staged an all-out strike in autumn 1977 was the TR7 Sprint. This had a 16-valve version of the 2-litre slant-four engine, like that used in the high-performance Dolomite Sprint saloon. This is the 16-valve engine in its saloon form.

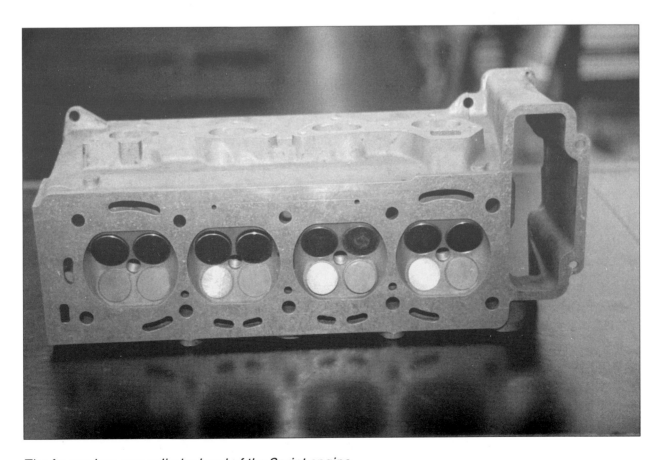

The four-valves-per-cylinder head of the Sprint engine.

Unique to the TR7 Sprint was the cast exhaust manifold seen here attached to a 16-valve cylinder head.

Under the bonnet of a genuine TR7 Sprint. Although sadly no longer in running condition, this car and its sister were rescued by TR7 parts specialists Rimmer Bros. Both were finished in Pageant Blue and both were registered in the SJW . . . S series which also contained Tony Pond's rallying TR7 Sprint.

This fascinating picture dates from August 1977 and appears to be a showroom mock-up for the forthcoming model range from Rover-Triumph. On the left is the Rover 3500 (actually launched in 1976); at the back is a TR7 (or possibly TR8) fixed-head; and on the right is the Lynx coupé, which was due to be launched in 1978.

Another picture from the same session in August 1977. The Lynx coupé is again on the right. The TR7/TR8 has a blacked-out tail panel and the early type of black wheel centre which had already been superseded in production. However, its front wheel appears to be an experimental alloy type, and the black stripe on the bodysides was never used in production.

The Lynx would have been a plushly-appointed 2+2 grand tourer. This is the interior of what was probably a styling mock-up.

THE CANLEY-BUILT TR7 FIXED-HEAD, 1978-1980

When TR7s started emerging from the Canley plant in the autumn of 1978. BL announced that the move of the production lines had allowed over 200 manufacturing and product improvements to be incorporated. This was rather difficult to believe from a cursory look at the 1979-model cars, which were recognisable only by their body-colour (instead of matt black) tail panels, by an extra bulge on the bonnet and by badging changes.

At the front, the TR7 decal badge had been replaced by another decal, which consisted of a styled laurel wreath, some nine inches in diameter, with the Triumph name emblazoned across it. The boot lid now bore "Triumph" in block letters on the left-hand side and "TR7" on the right, again in decal form. Badge colours – black, gold, or silver – depended on the main body colour.

Nevertheless, a closer inspection would reveal some of the improvements to the Canley-built TR7s. Though the interior remained otherwise unchanged, there was extra padding in the seats. An anti-chip coating on the body sills and in the wheelarches was designed to combat the corrosion which was experienced on earlier cars, and the motors which raised and lowered the headlamps were sealed against water and dirt in an attempt to eliminate one of the TR7's more common problems. Evidence of an attempt to tackle the electrical faults which had plagued earlier cars could be found in a reorganised fuse box, and a revised cooling system was expected to guard against the water loss problems which had so often resulted in cylinder head gasket failures. Even fuel economy received attention, and there was a hot air flap in the carburettor air intake to improve fuel atomisation in a cold engine.

With the transfer of production to Canley, the four-speed gearbox was finally dropped. The five-speed type was standardised for all markets, with the Borg Warner automatic as an option. In spite of this rationalisation, however, 1979-model North American TR7s were still quite different in other respects from their Rest-of-the-World counterparts. Only North American cars had a front spoiler, side running-lights and rear bumper overriders; and probably only North American models could be bought with air conditioning (although it might also have been fitted to cars for a small number of other hot-climate export markets).

Triumph's main priority for the 1980 model year was the introduction to the US market of the TR7 drophead and the V8-engined TR8s, and so it was not surprising that the four-cylinder fixed-head cars received very little in the way of improvements. As far as the home market was concerned, the only change for

1980 was that smart alloy wheels were now added to the options list; for North America, however, there were additional changes.

On 1980-model North American TR7s, the alloy wheels (identical to those planned for the TR8) were standard. Inside, the most obvious change was a new Moto-Lita steering wheel with three alloy spokes, but there were detail improvements, too: brighter warning lamps on the instrument panel, and a low-coolant indicator lamp. The twin steering column stalks also changed sides to conform to ISO standards, so that the lighting stalk was now on the left and the wipers on the right.

The TR7 was still primarily a North American export model, of course, and the production lines were kept busy in the second half of 1979 satisfying demand in that market. As a result, the 1980-model changes listed above were not seen on TR7s for the home market until January 1980 – and then only in a limited-edition model known as the Premium. There were just 400 examples of this, which was probably introduced mainly to give sales a fillip at a time when some customers were hanging back and waiting for the expected European announcement of the convertible TR7, which of course had been available in the USA for some months.

The Premium edition had no special identifying badges, but it was readily recognisable by its all-black paintwork, complemented by silver-grey side decals, alloy wheels with gold panelling, and the chin spoiler already seen on North American cars. Interior trim was in Navy or Tan, and was clearly fitted to test the market for these colours, which were slated to replace the standard Red and Green trims later on. The fabric sunroof optional on other TR7s was also fitted as standard. Less immediately apparent were the halogen headlamps, rear fog guard lamps mounted below the bumper, and the radio – normally an extra-cost option – which came as part of the Premium package.

Meanwhile, BL management had been looking at ways of rationalising production even further. One result was a decision to close the Canley assembly plant, and to transfer TR7 and TR8 production to the under-utilised Rover factory at Solihull. This time there was to be no sudden shut-down of production, however, and the lines were transferred progressively from April into the summer of 1980, so that there would be no hiatus in deliveries of new cars to dealers' showrooms. In practice, however, very few Solihull-built cars were delivered outside North America before the start of the 1981 model-year in summer 1980.

There was plenty of optimism when TR7 production restarted at Triumph's Canley plant. Note the reference to "new" TR7 in the banner stretched across the assembly line!

The quickest way of recognising one of the Canley-built cars when they were new was by means of the new wreath badge on the nose. In later years, some owners put these onto Speke-built cars, so they are not an infallible guide!

On 1980 models, smart alloy wheels became an option.

The Premium limited-edition model came in black with special side decals and a number of other special fitments. Just visible in this picture is the extra bonnet bulge which was another characteristic of the Canley-built cars.

In this rear view of a TR7 Premium, the value of that side decal is quite clear: it streamlines the car and makes it look longer. The styling sweep in the TR7's flanks had always been controversial, and was one feature which made the fixed-head car look short and dumpy. By the late 1970s Triumph was trying to find ways of de-emphasising it without having to go to the expense of changing the body tooling, and the near-horizontal lines of the decal did just that.

Curiously, the lettering on the decal recalled the badges on Speke-built cars rather than those on cars from Canley, where it was actually built!

A plastic chin spoiler was added during Canley production. Here it is on a 1979 UK-market drophead, albeit fitted with US-market side marker lamps.

BUYING A CANLEY-BUILT TR7 FIXED-HEAD

The build quality of the TR7s from Canley was vastly better than that of those made at Speke. As a result, these later cars are less likely to suffer from niggling faults than are the earlier examples. In addition, the improved corrosion protection makes them less prone to rust. One consequence is that these cars are in greater demand and command higher prices.

Just as with the Speke-built TR7s, however, most cars on the market are likely to have manual transmissions (in this case, five-speeds), because the automatics were relatively unpopular; the automatic box was also not available on Californian-specification TR7s at all. A good proportion of cars on offer will be fitted with the fabric sunroof, and some 50% of North American models are likely to have air conditioning.

Examples of the 400-strong Premium edition available on the home market in the early months of 1980 are not often seen for sale. It is quite possible that many of them have actually been resprayed in brighter colours by owners who found some difficulty in selling a black TR7 fixed-head! In any case, there is no indication at present that original-condition Premium cars command higher prices than ordinary fixed-head TR7s of the same vintage.

Triumph TR7

Latest in a long line of Triumph sports cars, the rakish TR7 sets a totally new standard in sports car design. Its distinctive aerodynamic body, built-in safety features and special 2 litre engine conform to the latest international legislation on safety and pollution as well as providing a luxurious and exhilarating motoring environment for two people. Instrumentation is superb and the controls layout has obviously been designed by people who know sports cars well. All in all, a superb example of Triumph engineering for a world of driving enthusiasts to enjoy.

Excerpt from J.R.T. brochure 1977.

THE SOLIHULL-BUILT TR7 FIXED-HEAD,
1980-1981

Sales of the fixed-head TR7 fell to their lowest-ever levels after the arrival of the TR7 drophead and, in North America, the TR8. Both these new models were so very much more exciting that customers for the much-maligned TR7 fixed-head were few and far between. The fact that the fixed-head actually cost *more* than the more desirable open car undoubtedly depressed sales further: clearly BL was trying hard to phase out the fixed-head car altogether. Nevertheless, some 1,500 fixed-head TR7s *were* built at Solihull after production was transferred there from Canley in the spring and summer of 1980.

Like Canley, Solihull put its own stamp on the TR7. Cars built at the Rover works had a small plastic wreath badge on the nose instead of the large decal, and "2.0-litre" badges on the front wings, presumably to distinguish them more readily from the V8-engined models. They were also offered in a new range of colours, which included metallics for the first time (the metallics planned for 1978 models had not in practice been available because of the Speke strike), and they had plaid upholstery in Navy or Tan instead of the Green and Red options of the earlier cars. Dashboards were now moulded in blue or tan to match the upholstery, and this represented a major visual improvement over the stark black dashboards of Speke and Canley-built cars. Less obvious, though at least as important in terms of sales appeal, was the fact that the 1981-model TR7s needed less servicing than earlier examples, and were thus cheaper to run.

In order to simplify production at Solihull, specification differences between North American cars and those destined for other markets had been eliminated as far as possible. Thus, alloy wheels and the large plastic chin spoiler became standard on all cars, as did the "reversed" ISO control stalks on the steering column, the brighter dashboard warning lamps and the low-coolant-level indicator. Nevertheless, Federal models had a new design of steering wheel with a circular centre boss which was not fitted to other cars. For the home market, the fabric sunroof, fog guard lamps below the rear bumper, and a radio all became standard. At the same time, the poor-selling automatic version of the TR7 ceased to be available on the home market.

Where both Speke and Canley had been obliged to cope with two different versions of the TR7 for North America – the "49-State" version and the Californian version with its different exhaust emissions control equipment – Solihull introduced a further production rationalisation. All Solihull-built North American TR7s had a Lucas-Bosch fuel injection system in place of the Zenith-Stromberg carburettors on earlier cars, and this met emissions requirements in all States. Both manual and automatic fuel-injected cars had higher axle ratios (borrowed from the Rover SD1 whose axle the TR7 used) to give improved fuel economy: the manual cars went up to 3.45:1 while the automatics went to 3.08:1. Outside the North American market, however, the carburetted engine and original axle ratio remained standard.

The 1981 cars would be the only ones built at Solihull, for BL decided to bring production of its last remaining sports cars to a halt in October of that year.

The plastic bonnet badge fitted only to Solihull-built cars.

Also found only on TR7s built at Solihull was this "2.0 LITRE" badge, fitted to the front wings.

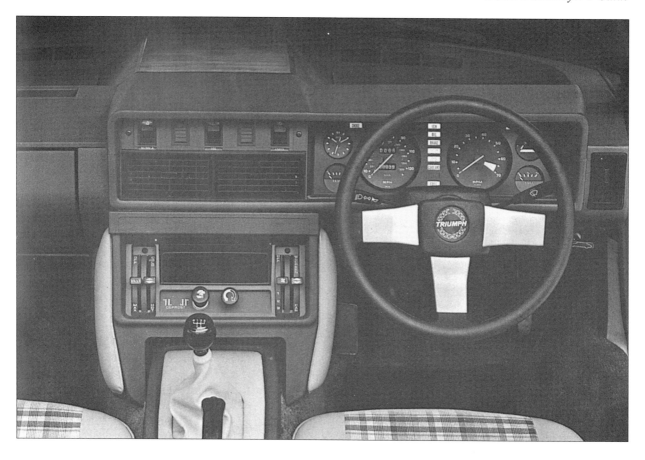

Solihull-built cars also had a more attractive looking steering wheel.

The combination of folding sunroof and the latest interior materials made the fixed-head car a much more attractive proposition than it had been before; but the drophead proved far more popular.

The folding sunroof was standard on fixed-head TR7s built at Solihull. In this picture, the small plastic bonnet badge which characterised cars built at the Rover factory is also in evidence.

B U Y I N G   A   S O L I H U L L - B U I L T   T R 7
F I X E D - H E A D

Quality of the Solihull-built TR7s was every bit as good as that of the Canley cars – indeed, some people insist that it was even better. However, it will not prove easy to find a Solihull fixed-head TR7 for sale because relatively few were built. One day, the combination of low build volumes with high build quality might cause prices of these cars to rise above those of other TR7 fixed-head models; for the moment, however, the fact that the fixed-head TR7 is generally perceived as the least desirable of the TR7/TR8 range is keeping prices low. Few sellers will expect to realise significantly more for a Solihull TR7 fixed-head than they would for an equivalent Canley car.

Metallic paints were popular options on the Solihull cars, and were probably applied to a majority of these late fixed-head TR7s for all markets. Fabric sunroofs were more or less a universal fitment (although there might have been some export markets which took cars without them). On the home market, all cars on offer will have the five-speed manual gearbox, although automatics continued to be built for North America.

As far as the North American market is concerned, the major distinguishing features of the Solihull-built 1981-model cars are the fuel injected engine and the raised final drive gearing. The fuel injection itself is generally trouble-free, and it brings with it an increase in engine flexibility, particularly as compared to the emissions-controlled carburettor cars. Even with the higher gearing, this means that there is no loss of driveability, while top speeds are up and fuel consumption is down.

Overall, the 1981-model TR7 fixed-heads from Solihull are a very good buy for someone who is content with the closed bodywork configuration on a TR7.

THE TR7 DROPHEAD, 1979-1981

Once production was under way again at Canley, Triumph forged ahead with the introduction of new variants, and the long-awaited drophead body shell was finally ready for introduction by the summer of 1979. As had been the case when the TR7 was first introduced, the priority was given to North American sales, and the TR7 drophead went on sale in that market on 1st July 1979, a full nine months before home market and other versions became available.

These 1980-model dropheads were built at Canley, as were the very first cars for the home market, where the drophead was announced in March 1980 (the European launch was at the Brussels Motor Show in January 1980). However, the transfer of production to Solihull began immediately after the drophead's UK launch, with the result that Canley built relatively few dropheads for the home market. The majority of home market dropheads and all 1981-season North American cars were built at Solihull.

As far as recognition features are concerned, the Canley and Solihull versions of the TR7 drophead are simple to distinguish. All Canley-built cars had the large wreath decal on the nose, whereas all Solihull-built cars had the smaller plastic badge, in exactly the same way as their fixed-head counterparts. However, all dropheads, whether built at Canley or at Solihull, have Navy or Tan plaid upholstery; Canley-built fixed-head cars, of course, had Red or Green plaid upholstery. Similarly, all dropheads, whether from Canley or Solihull and whether for the US or other markets, had a plastic chin spoiler; Canley-built fixed-head cars for the home market did not.

There was little doubt – and contemporary press reports echo this view – that the drophead body was very much better looking than the fixed-head version. It also made the TR7 *look* more like the sports car it had always been intended to be, with the result that many people who had claimed the TR range had died with the TR6 began to have second thoughts. Yet, as the story of the car's development earlier in this book shows, the basic lines of the fixed-head's body had been retained in their entirety.

Under the skin, however, there had been a great deal of structural re-engineering. The body had been reinforced with box-section bracing behind the rear seats, linking the door shut pillars, while the rear quarter-panels were extended downwards to anchor into the side sills. At the front, both the cross-member and the suspension turrets were reinforced to prevent scuttle shake; but particularly interesting was the way in which the front bumper was used to disguise what was effectively a harmonic damper. Weights were concealed in the outer ends of the bumper, and the inner steel armature pivoted in the middle. The idea

was not new: the Wilmot-Breeden harmonic stabiliser, which relied on similar principles, had been fitted to many cars of the 1930s in order to counteract front axle tramp!

The interior of the dropheads was unchanged from that of the fixed-head cars, except that the latter's moulded rear parcels shelf was replaced by a well into which the convertible top retracted when not in use. A neat tonneau cover then concealed the hood, and contributed to the smooth rear deck line of the open car. The top itself was double-skinned and fitted neatly and tightly to the tops of the window frames. An unzippable rear window was provided, along with fixed panes in the rear three-quarter panels, and the hood was attached when erect by press-studs on the rear body and by two large catches on the windscreen header rail. Its raked rear screen encroached more upon the rear deck than did the rear window of the fixed-head cars, and so the fuel filler had been made smaller in order to leave sufficient room.

Although the loss of the metal roof removed some weight, the hood frame and canvas and the additional reinforcement in the drophead body shell meant that the whole car weighed about two hundredweights more than its fixed-head counterpart. Nevertheless, performance was not adversely affected, and in fact the top speed was a little higher, due no doubt to the better aerodynamic shape.

For 1980 in North America, there were two distinct variants of the car; a 49-State Federal version with twin carburettors and 86bhp, and a single-carburettor 76bhp California-only model. Transmission options were the five-speed manual or three-speed automatic, although the latter was not available on the lower-powered Californian cars. Air conditioning, a radio and metallic paint were all options. Outside the USA, the few 1980 models built of course all had the usual twin-carburettor engine with 105bhp, and by far the majority of these had the five-speed gearbox. Extra-cost options on the home market were the automatic transmission, metallic paint, and alloy wheels, while a full tonneau cover could be bought as an accessory through BL's Unipart division.

For 1981, the Solihull-built cars for North America all had the fuel-injected engine, with either the five-speed or automatic gearboxes driving to the same higher-ratio rear axles as were fitted to the contemporary fixed-head TR7s. Solihull-built 1981-model cars for the home market again had the twin-carburettor engine but could be had only with the five-speed gearbox. Alloy wheels were now standard. The final dropheads, like the final fixed-heads, were built in October 1981.

The drophead was produced in two special-edition

variants for North America, both limited in numbers in the usual way. These were called the Victory edition and the Spider – the latter an all-black car with subtle side striping, a decal bearing its name just ahead of the rear wheel arches, and alloy wheels.

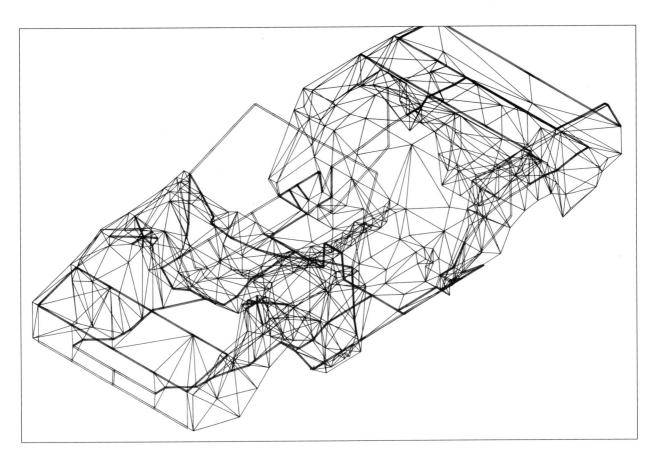

BL released this Finite Element Computer Analysis Diagram to the press when it announced the drophead TR7 in the UK. It was intended to stress the role of computer-aided design in developing the drophead body from the original fixed-head type.

The side view of the drophead in open form was so much more attractive than that of the fixed-head car that it was hard to credit their common ancestry.

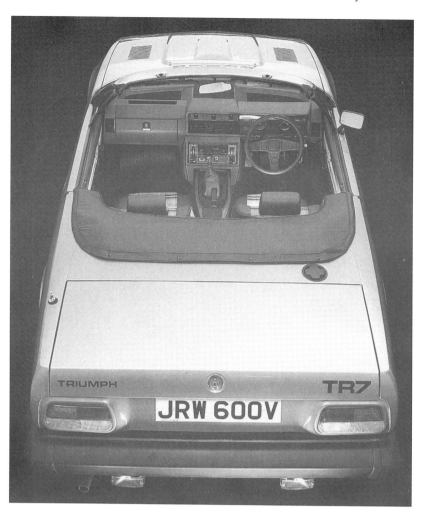

This early press-release picture of a right-hand-drive TR7 drophead emphasised the open-air nature of the car.

These small edge-protectors on the boot lid were not fitted to JRW 600 V in publicity pictures taken of the car, but all production cars had them.

The three-window convertible top fitted neatly and looked much more stylish than the metal roof of the fixed-head car.

From this angle, the open car looked long and sleek, although it was in fact no longer than the fixed-head type.

Drophead bodies, mostly for left-hand-drive models, await their turn on the assembly lines at Solihull. Just visible on the right are the Rover SD1 saloons which were assembled in the same plant.

A crucial stage in TR7/TR8 drophead assembly: the painted body is brought into the plant on an overhead conveyor and is lowered onto the running gear which has been built up on the assembly line.

A TR7 or TR8 drophead, now lacking only final details such as badges, is tested on the rolling road at Solihull.

The longer Federal specification bumpers, fitted since the beginning of TR7 production, can be seen in this picture of a Canley-built Federal drophead. The rubbing-strip on the body side is a neat touch, but not original. (Picture by courtesy of Dave Destler at British Car magazine.)

The interior of a 1980-model Federal TR7 drophead. This type of steering wheel was unique to Federal models and to the TR8. (Picture by courtesy of Dave Destler at British Car magazine.)

A 1981-model Federal TR7 drophead. Note the wheel trims, which were never made available on UK models.

BUYING A TR7 DROPHEAD

There is little doubt that the drophead body greatly enhanced the TR7's appeal. It is good-looking where the fixed-head was ungainly, and above all it looks like a sports car and offers open air motoring like that which was essential to the character of the TR2 to TR6 models which went before it. The body is rigid, and suffers from none of the shake familiar from earlier open TRs, and the low seating position, high waistline, and large screen ensure that wind buffeting is minimised when motoring with the top down. The soft top itself is easy to erect – though if it is not lowered properly, the tonneau cover can be difficult to fit – and insulates the occupants well against both weather and wind-noise. Rearward visibility is also much better than in the fixed-head, thanks to the rear three-quarter windows, although the plastic material used for the hood windows does distort vision a little.

As far as the home market cars are concerned, the differences between the Canley and Solihull versions of the car are too small to make a persuasive case for buying one rather than the other. Both are well-made, and neither suffers from the sort of faults which let the early fixed-head cars down so badly. In years to come, the greater rarity of the Canley models might increase their desirability and therefore prices, but at present the determining factor when buying remains condition. Automatic transmission – available only on 1980-season cars – is very rare, and the home market TR7 drophead is effectively available with the five-speed gearbox only.

The picture is rather different for the North American cars, as there are major differences between 1980 and 1981 models. On the whole, the 1981 (Solihull) models with their fuel-injected engines are preferable – particularly in California, where the single-carburettor 1980 models are very much underpowered. Just as with the fixed-head cars, automatic transmission was not popular, and consequently anyone who is determined to have an automatic TR7 drophead may have to look quite hard. Performance, obviously, is not the strong point of one of these cars. Air conditioning was probably specified less commonly than on the fixed-heads, but is nevertheless often found. The alloy wheel option was also fairly common on 1980 models (alloy wheels were standard for 1981). The Spider and Victory special editions are unlikely at present to command prices significantly higher than those of the "ordinary" cars, although the differential may increase as enthusiast interest in the TR7 grows.

THE TR8

As had been the case with the TR7, British Leyland decided to launch the TR8 on the US market before making it available elsewhere. It was for this reason that the pre-production TR8s – fixed-head cars built at Speke in 1977 just before the strike – were shipped out to the USA for trials and assessment work. As far as it is possible to tell, BL's intention was to introduce the TR8 in the US market in the spring of 1978, but the Speke difficulties of course put paid to that idea. Nevertheless, the TR8 did eventually go on sale in the USA, where it was introduced on 1st May 1980 – some two years later than originally planned.

As part of the promotional campaign for the TR8, BL was planning to use the car in competitions. Development of the "works" TR8s was carried out at Abingdon, and was therefore unaffected by the Speke strike, with the result that the cars were ready for homologation on schedule in the early spring of 1978. Even though the competitions programme was now out of phase with the production programme, BL decided that it should go ahead, and on 1st April 1978, the TR8 was homologated for rallying under the name of "TR7 V8". A number of the existing works rally TR7s, now fitted with 3½-litre V8 engines, were seen in competition that season.

BL had every intention of making TR8s available outside the USA, of course, and development work focused on cars for other markets as soon as the US models had gone on sale in 1980. A number of pre-production right-hand-drive dropheads had been built and were under assessment by the summer of 1981, but both TR7 and TR8 were taken out of production that autumn and, as far as buyers outside the USA were concerned, the TR8 never existed.

Parallel development from the beginning had ensured that there were relatively few specification differences between the TR8 and the TR7, which of course simplified the production process and saved costs. The major difference between the two cars was the engine itself, and all the specification changes which were not purely cosmetic were associated with that. As the Rover engine was heavier than the slant-four, different springs and dampers had to be fitted, and the battery was moved from engine bay to boot in order to give a more even weight distribution. Even so, the TR8 was relatively nose-heavy, with a 57%/43% weight distribution between front and rear. That was one reason why power-assisted steering was made standard on the car.

Weight was one major consideration; the other was the engine's greater power and torque. With the same rear axle ratio as the TR7, the TR8 would have offered neck-snapping acceleration but little more in the way of top speed, and fuel economy would have been poor

for a two-seater sports car. As a result, the car was given the SD1's taller 3.08 axle ratio, which was low enough to ensure that acceleration was much better than the TR7's, and high enough to keep fuel consumption within reasonable bounds.

The engine itself, of course, was already well-known and widely respected. It was not in fact originally designed by Rover, but had been bought from General Motors in America, who had used it in their "compact" Buick, Oldsmobile and Pontiac models in the early 1960s. Redeveloped to suit British manufacturing methods and motoring conditions, it first appeared in Rover saloons in 1967 and had been a staple of the company's range ever since. The same engine was also supplied to specialist car builders like Morgan and TVR, and had become a favourite as a conversion unit among performance enthusiasts.

The TR8 engine was different in many respects from the original 1967 Rover engine, although it still had the same bore and stroke dimensions and the same overhead valve layout with a single chain-driven camshaft operating the valves by pushrods and rockers. As the expansion rates of the aluminium alloy cylinder block and heads differed from those of the steel valve gear, hydraulic tappets were used in the valve-train to maintain clearances (and therefore noise levels) at a constant level.

For the SD1 and TR8 applications of the engine, more top-end power had been obtained by fitting bigger valves and redesigned hydraulic tappets which allowed the engine to rev higher. As first tried in the TR8 prototypes in 1973, the V8 was fitted with twin SU carburettors and low-line air filter boxes which were probably borrowed from the Australian-specification Range Rover version of the engine. Even this, however, did not keep the engine low enough to fit under the car's sloping bonnet, and at least one prototype had rather crude bulges added to its bonnet to give enough clearance for the SUs. For production engines with carburettors, flatter Zenith-Strombergs replaced the SUs; and a fuel-injected version of the engine was also developed in the later 1970s to cope with North American emissions control requirements.

There were four major variants of the TR8, if the works rally team cars (technically TR7 V8s) are discounted. These were the 1977 Speke-built fixed-head cars, the 1980-model US market dropheads, the 1981-model US market dropheads, and the 1981 pre-production "rest of the world" dropheads. Of these, the 1977 Speke-built pre-production cars are undoubtedly the odd men out. Exactly how many there were is impossible to establish, but one source claims that there were 88 built, of which 78 were shipped to the USA for pre-launch trials. As the cars were still theoretically secret during these trials, they had no identifying badges. The Speke TR8s are, of course, the only fixed-head cars built, as all production TR8s were dropheads. Unlike other TR8s, they also had upholstery in ribbed cord rather than tartan cloth; and, where production cars have the battery exposed in the boot, the pre-production cars had it under a clip-down lid.

The production TR8 dropheads for the 1980 season were built at Canley. Distinctive badging made them easy to distinguish from their four-cylinder TR7 contemporaries at a glance: instead of a wreath badge on the nose, there was a large "TR8" decal and, at the rear, the boot lid had a "Triumph" decal on the left and a "TR8" decal on the right. The shaded lettering used for both of these badges made them quite different from the contemporary TR7 badges, which had solid lettering. In addition, there was the Triumph name in a wreath above "3.5-litre" on each front wing, just ahead of the doors. All these badges had been designed in the USA, and were expected to have special appeal to US customers.

The interior of the 1980-model TR8s was broadly similar to that of their TR7 contemporaries. The major differences were that the TR8s had a different rev counter to suit the eight-cylinder engine, and "TR8" badging on the glove-box lid. Radios were extra-cost options (but very few cars can have been supplied without them), and all cars had the 85mph speedometer required by Federal law. In fact, the TR8s were capable of considerably higher speeds than that.

The 1980 TR8s were produced in two distinct versions, as the exhaust emission rules in California remained stricter than those for the other 49 states. In "49-state" form, the TR8's engine had twin Stromberg carburettors and put out 133bhp at 5,000rpm with 174 lb/ft of torque at 3,000rpm. This was said to make the cars good for 120mph and 0–60mph in 8.4 seconds. Californian TR8s for 1980, however, had fuel-injected engines, with a three-way catalyst and an oxygen sensor in the exhaust. The injection system itself was a Bosch K-Jetronic type, produced under licence by Lucas, and the Californian cars had 137bhp at 5,000rpm and 165 lb/ft of torque at 3,200rpm. They were thus slightly more powerful and rather less torquey than the "49-state" TR8s.

Both "49-state" and Californian cars could be had with five-speed manual or three-speed automatic transmissions, although the latter was very much less common. Both types had the same 3.08:1 rear axle; both had power-assisted steering as standard; and both had alloy wheels. Both, of course, also had twin exhaust pipes, which both helped to differentiate them visually from TR7 dropheads and hinted at the higher performance available.

When production was transferred from Canley to Solihull in the middle of 1980, the TR8's specification was altered slightly, and the Solihull-built cars became the 1981 models. There was just one major change: the Californian fuel-injected engine with its exhaust catalyst was now standardised for all states, in anticipation of 1983 legislation which would require exhaust catalysts on all new cars sold in the USA. Fifth gear on manual-transmission models was also raised to improve fuel economy at cruising speeds. And metallic paints were almost universal on 1981 cars.

The TR8 was never a catalogued model in any country outside the USA, and the only "home market" models ever built were a handful of prototypes and pre-production dropheads made in 1981 at Solihull. The exact quantity built is in doubt: records from the Solihull plant (see under "Production Figures" at the end of this book) suggest that nine were built at the end of 1980 and a further nine in 1981. However, other convincing documentation suggests that a total of 36 home-market TR8 dropheads came off the lines. The latter figure seems more plausible, but cannot yet be verified. Neither figure appears to include the actual prototypes, which might have added as many as six to the total.

When TR8 production was aborted in October 1981, the Solihull factory was just gearing up for the model's UK launch. According to one (plausible) version of the story, the press demonstrator fleet had been completed and the first showroom models had started to come off the lines. The factory registered 22 cars (mainly with W-suffix registrations) and auctioned them off in October 1981, and a further 14 were released to Triumph dealers and were eventually sold on.

The "home market" cars of course had right-hand drive, and they had carburetted engines to the same specification as the 1982-model SD1s (which would not actually be released until January 1982). That meant that they had a 9.35:1 compression ratio, 157bhp at 5,250rpm and 198 lbs/ft of torque at 2,500rpm. Performance figures were never released by the factory, but the engine specification makes it clear that these cars would have been considerably faster than the US-specification models. At a guess, 135mph would have been possible, with 0–60mph times of under 8 seconds.

The home-market TR8s had the alloy road wheels and three-spoke Moto-Lita steering wheel of their US counterparts. They also had the ribbed cord seats seen on the 1977 Speke-built TR8 coupé prototypes rather than the tartan cloth seats standard on US-market TR8s. Some cars left the factory with no badging on the front wings, while others had the US-type decal badges. Not every "home-market" TR8 which escaped from the factory had the same specification: some were used for development purposes, and at least one car was used for trim and cosmetic development work.

Not many UK-specification TR8s were built. This one, now owned by Triumph parts specialists Rimmer Bros, was originally white but has been resprayed in an attractive dark metallic grey. The TR8 looked very similar to the TR7 from most angles.

The badges on the front wings, clearly visible from a distance, were one feature which readily distinguished a TR8 from a TR7.

Other badging also made clear what the car was. On the back was this . . .

. . . and the front view revealed this. Only the "TR8" decal was present on 1980 models; the plastic Triumph badge was left off.

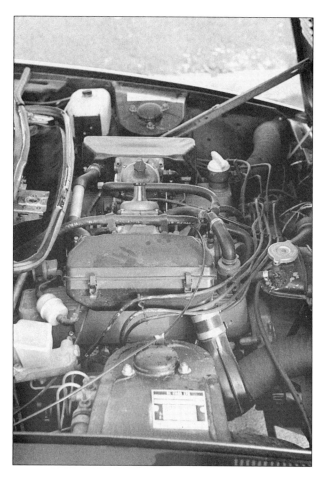

The Rover V8 engine fitted snugly under the bonnet of the TR7 to produce a TR8. This picture shows the V8 engine in one of the genuine UK-specification TR8s built over the summer of 1981.

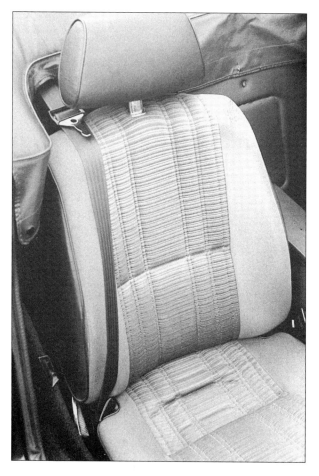

Upholstery material differed from that used in TR7s, as this picture of RDU 35 W shows.

TR8s for all markets had an attractive three-spoke steering wheel.

The shaded lettering of the decal badges was seen in miniature on the glove-box lid.

B U Y I N G   A   T R 8

In many ways, the convertible eight-cylinder car was the best of the whole TR7/TR8 range. It had on its side both the good looks of the open body and the performance of the V8 engine. The fact that it lasted only two seasons in production makes it one of the rarer variants. Rarer still are the 1977 TR8 pre-production fixed-head cars, although their desirability is questionable because they are closed rather than open cars and are likely to suffer from all the familiar Speke build quality problems. Rarest of all are the home-market TR8s.

This means that there will be very few TR8s on the market at any one time; and in the UK, the market is effectively non-existent because the cars so rarely come up for sale. Even in the USA, the numbers of those for sale are restricted by the fact that many were bought as "investments" after the end of production had been announced. As values have not yet escalated dramatically, these cars are unlikely to appear on the market yet. When values do increase, however (or if owners get tired of waiting), expect several unused or very low-mileage TR8 convertibles to come on to the market!

Most owners of TR8 convertibles are aware of the car's potential "collectible" status, with the result that they generally pamper their cars more than do owners of TR7s of similar vintage. It therefore follows that examples for sale are not likely to be cheap unless there is something very obviously wrong with them. However, as values have remained fairly static, there is not a great deal of difference between the going rate for a well-used car and that for a well-kept one. Anyone looking for a TR8 convertible should therefore shop around carefully, be patient, and not buy the first example to become available unless it really is a good car.

As far as performance is concerned, there is little to distinguish the later (and more numerous) fuel-injected cars from the carburetted models. More torque lower down the rev range makes the 1980 "49-state" cars a little more responsive, but the TR8 has so much torque in any case that the difference is all but academic. The higher fifth gear ratio on 1981 models permits more relaxed cruising and slightly better fuel economy. Automatic-transmission cars are relatively rare and are slightly more thirsty than the five-speed models.

In most respects except for their engines, the convertible TR8s share the strengths and weaknesses of the contemporary TR7 convertibles. Their greater performance is likely to make the ride and handling deficiencies caused by worn suspension bushes more readily apparent than in the four-cylinder cars, however. Unlike the TR7s, the TR8s of course had power-assisted steering, and this does impart a rather different "feel" to the car. Some owners consider that the degree of power assistance is too great to provide the sporting handling expected of a sports car. On the positive side, however, the PAS system does appear to be largely trouble-free in service.

As for the TR8's engine, this is a robust piece of machinery which should be good for at least 100,000 miles without major overhaul. Noise from the top end is to be expected in high-mileage examples and may indicate a worn camshaft, but in a low mileage engine it will probably be caused by oil sludge blocking the valves in the hydraulic tappets which are designed to take up a certain amount of wear and to keep the engine running quietly. This usually means that regular oil changes have not been carried out: the manufacturers recommended an oil and filter change every 7,500 miles, but many experienced owners believe there are advantages in changing the oil more regularly than that.

Acceleration in a TR8 is very rapid, and bears comparison with many newer and larger-engined cars. High-speed cruising is very relaxed, particularly in the overdrive fifth of the manual gearbox. Nevertheless, fuel consumption, at around 15–16 miles per U.S. gallon (around 18mpg by Imperial measures), is not one of the car's stronger suits. Worth noting is that the fuel-injected cars are more likely to need specialist attention, whereas the carburetted cars may be easier for the DIY owner to work on.

Options for the US market were few in number. Air conditioning was an extra-cost option on 1980 models, but in practice relatively few cars seem to have been supplied without it.

Home-market TR8s of course bring their own problems for the buyer. The first thing to remember is that genuine home market TR8s are very rare indeed and are unlikely to come up for sale very often. Prices, traditionally, have been high.

Very few of these cars are used on the road regularly, and therefore low mileages are common. Rarity value also ensures that they are carefully looked after, so condition is not likely to prove either a deterrent or a bargaining point if and when one comes up for sale.

The biggest danger for the enthusiastic buyer is that an unscrupulous seller will offer for sale as a genuine TR8 what is actually a converted TR7. Such conversions are not hard to achieve and several competent and reputable engineering companies have specialised in putting V8 engines into TR7s – though none of them would deliberately attempt to mislead a buyer into believing one of their conversions to be a factory-built TR8.

Unfortunately, there is no infallible way of telling whether or not a car is a genuine factory-built home market TR8. Documents can go a long way towards proving authenticity, commission numbers and engine numbers are a fairly reliable guide (but are not impossible to alter), but it is not easy to prove that a 1980 car with a TR7 commission number and SD1 engine was not a factory prototype. In cases of doubt, a check with the build records held by the British Motor Industry Heritage Trust should help to establish authenticity.

V8 power with Nick and Heather King in their TR7.
Photo Jeremy Belcher (Courtesy of TR Register).

MODIFICATIONS

* One of the more popular TR7 conversions has been to update the engine to 16-valve "Sprint" specification, along the lines of the TR7 Sprint which would have gone into production if the 1977 Speke strike had not intervened. This adds between 25bhp and 40bhp to the peak power of the 2-litre engine. Both complete engines and conversion parts are readily available, as the 16-valve engine was used in the Dolomite Sprint saloon. Some Triumph specialists offer a full conversion kit off the shelf. All TR7 Sprint conversions offer higher performance than is available from the standard car, and must therefore be matched by an uprated braking system. Home conversions without uprated brakes are potentially dangerous and should be avoided.

* It is possible to fit a Dolomite overdrive to a four-speed TR7. Some owners have seen this as a way of improving fuel consumption which is more cost-effective than fitting a five-speed gearbox, which of course has to be accompanied by the so-called heavy-duty rear axle.

* Many TR7s have been converted to TR8 specification, a relatively simple conversion because the TR7's bodyshell was designed to accept the Rover V8 engine. All such conversions need uprated brakes, of course, and they should also have either the five-speed manual transmission or three-speed automatic, as the four-speed TR7 gearbox is not strong enough to take the torque of the V8. Specialists have for some years offered complete kits to enable enthusiasts to undertake the conversion at home. Others – notably Grinnall –

have carried out further development on the car to produce a high-performance vehicle which is rather more than a simple V8 conversion of a TR7.

* During the production life of the TR7 drophead, Lenham introduced a GRP hard-top for the car. The manufacturing and distribution rights for this removable hard-top later passed to Rimmer Bros.

* Body styling kits consisting of add-on GRP panels were first introduced in 1984 by Triumph specialists Rimmer Bros. The Rimmer Bros kit has been produced in three different versions and has spawned other body kits from other manufacturers.

* More ambitious than the body kits available elsewhere for the TR7 was the full conversion kit produced for the TR7 by Eurosports UK, of Rayleigh in Essex. The TR40 kit, as it was known, was introduced towards the end of 1990, when the TR7 fixed-heads for which it was intended were available cheaply. The kit consisted of new GRP front and rear ends, door skins and other panels which, when supplemented by new wheels and tyres and lowered suspension, made a car which broadly resembled the Ferrari F40 (from which the kit's name was clearly derived).

* Although the TR7 and TR8 are strictly two-seater cars, a proper jig-built 2+2 conversion has been engineered by S and S Preparations of Ramsbottom, Lancashire, and figures in their extensive list of upgrades and modifications for the cars.

Hurley Motor Engineering offered to convert TR7s to Ford V6 power in the early 1980s. The conversion did not become popular, however.

The Hurley V6 conversion was rebadged as a TRH.

The Lenham hard top for the TR7 drophead could be fitted with a tilt-and-remove glass sunroof, as on this example.

The fixed-head TR7 ancestry of the Eurosport TR40 was not immediately apparent.

Conversions from TR7 to TR8 are popular. These pictures show two different approaches, one using twin SU carburettors and the other using an American Holley.

A professionally converted TR7 drophead, now fitted with a Rover V8 engine and a number of other non-standard extras.

SPECIFICATIONS

North American TR7 – 1975 to 1981

Engine: 1,988cc (90.3mm bore x 78mm stroke) four-cylinder with single overhead camshaft; 8.0:1 compression ratio, two Zenith-Stromberg carburettors; 92bhp at 5,000rpm (1980 models: 86bhp) and 115 lbs/ft of torque at 3,500rpm. Californian cars have a single Zenith-Stromberg carburettor and 76bhp. 1981 models for all states have Bosch fuel injection.

Transmission: Single-dry-plate clutch. Four-speed and reverse gearbox; ratios 2.65:1, 1.78:1, 1.25:1, 1.00:1, reverse 3.05:1: axle ratio 3.63:1. Optional Borg Warner type 65 three-speed and reverse automatic gearbox: ratios 2.39:1, 1.45:1, 1.00:1, reverse 2.09:1; axle ratio 3.27:1 to 1980 and 3.08:1 for 1981. Five-speed and reverse LT77 manual gearbox replaced four-speed in summer 1976; ratios 3.32:1, 2.09:1, 1.40:1, 1.00:1, 0.83:1, reverse 3.43:1; axle ratio 3.9:1 to 1980 and 3.45:1 for 1981.

Steering, suspension and brakes: Rack-and-pinion steering. Independent front suspension with coil springs, MacPherson struts and anti-roll bar. Live rear axle with coil springs, radius arms, anti-roll bar and telescopic dampers. Servo-assisted brakes with 9.7in discs on the front wheels and 8in x 1.5in drums on the rear wheels; 9in x 1.75in drums on five-speed models. 175/70 SR 13 tyres on four-speed and automatic models; 185/70 HR 13 tyres on five-speed models.

Dimensions: Overall length 13ft 8.5in; overall width 5ft 6.2in; overall height 4ft 1.9in; wheelbase 7ft 1in; front track 4ft 7.5in; rear track 4ft 7.3in. Unladen weight 2,241 lbs (four-speed fixed-head); 2,355 lbs (five-speed fixed-head); 2,387 lbs (five-speed drophead).

"Rest of the World" TR7 – 1976 to 1981

(As for North American model, except:)
Engine: 9.25:1 compression ratio, two SU carburettors; 105bhp at 5,500rpm and 119 lbs/ft of torque at 3,500rpm.

Transmission: Five-speed gearbox optional only between September and December 1976; standard from summer/autumn 1978.

Dimensions: Overall length 13ft 4.1in. Unladen weight 2,205 lbs (four-speed fixed-head); 2,311 lbs (five-speed fixed-head); 2,356 lbs (five-speed drophead).

North American TR8 – 1980 to 1981

(As for North American TR7, except:)
Engine: 3,528cc (88.9mm bore x 71.1mm stroke) V8-cylinder with overhead valves; 8.1:1 compression ratio, two Stromberg carburettors; 133bhp at 5,000rpm and 174 lbs/ft of torque at 3,000rpm. 1980 Californian models and all 1981 cars have Lucas fuel injection, 137bhp at 5,000rpm and 168 lbs/ft of torque at 3,250rpm.

Transmission: Five-speed LT77 manual gearbox standard; Borg Warner type 65 three-speed automatic optional; axle ratio 3.08:1 for both types.

Steering and brakes: Power-assisted steering. 9in x 1.75in rear drums.

Dimensions: Unladen weight 2,565 lbs.

VEHICLE IDENTIFICATION

TR7s and TR8s built at Speke and Canley have their *vehicle identification numbers* on a plate attached to the left-hand door shut face, below the lock striker plate. Solihull-built cars have the number on a plate attached to the right-hand front suspension turret.

Before summer 1979, these numbers consisted of between one and five digits, and had three-letter prefixes, as follows (note that some sequences were used for pre-production cars only):

a) Speke-built cars

ACG	1976—1978 "Rest of the Word" TR7 fixed-head
ACH	1977 TR7 Sprint fixed-head
ACL	1975–1976 North American TR7 fixed-head
ACN	1977–1978 North American TR8 fixed-head
ACT	1978 North American TR7 drophead
ACW	1977–1978 North American fixed-head

b) Canley-built cars, 1978–1979

TCG	"Rest of the World" TR7 fixed-head
TCN	North American TR8 fixed-head
TCT	North American TR7 drophead
TCV	North American TR8 drophead
TCW	North American TR7 fixed-head

(Typical identification numbers in these sequences would be ACG 4500 and TCT 105825.)

From summer 1979, standardised Vehicle Identification Numbers (VINs) were used. The six-digit numbers are preceded by eight-digit prefix codes, which break down as follows:

1.	T	Triumph	(Make)
2.	P	TR7 or TR8	(Model range)
3.	A	UK or European market	(Market)
	J	Japanese market	
	K	Australian market	
	L	Canadian market	
	V	US Federal market (49 states)	
	Z	Californian market	
4.	D	Drophead	(Body style)
	E	Fixed-head	
5.	J	Four-cylinder engine (TR7)	
	V	Eight-cylinder engine (TR8)	
6.	3	RHD, automatic	(Steering and
	4	LHD, automatic	transmission)
	7	RHD, manual	
	8	LHD, manual	
7.	A	1980-1981 model (Rest of the World)	(Major
	B	1981 models (US market)	specification
	C	1982 models (US market)	changes)

8.	A	Solihull	(Assembly
	T	Canley	plant)

From May 1981, these prefixes were increased to eleven digits by the addition of SAT (the World Manufacturer Code identifying British Leyland's Triumph marque) in front of the eight-digit codes already in use.

Typical numbers with these sequences would be TPAEJ7AT 215050 and SATTPVDV8BA 406010.

Engine numbers on TR7s will be found adjacent to the carburettors. On TR8s, the engine number is on a ledge beside the dipstick on the left-hand cylinder bank.

The type identifiers for TR7 engines are as follows:

CG	European specification
CL	Australian, Canadian and North American engines to 1977
CK	North American engines with fuel injection
CV	Japanese and North American engines from 1977

The type identifiers for TR8 engines are as follows:

10E	Federal specification, manual transmission
11E	Federal specification, automatic transmission
12E	Californian specification, manual transmission
13E	Californian specification, automatic transmission
14E	Fuel injected versions, manual transmission
15E	Fuel injected versions, automatic transmission

Gearbox numbers will be found on the top face of the main casing, just behind the bellhousing. Identifying numbers for TR7 gearboxes are:

CG	Manual, four-speed
CL	Manual, five-speed (Canley and Solihull assembly)
14A.CL	Manual, five-speed (Speke assembly)
027M	Automatic, Borg Warner type 65, carburettor engines
063J	Automatic, Borg Warner type 65, fuel injected engines
076N	Automatic, Borg Warner type 66, carburettor engines
077K	Automatic, Borg Warner type 66, fuel injected engines

Identifying numbers for TR8 gearboxes are:

15A	Manual, five-speed
051N	Automatic

Finally, rear axles also bear an identifying number, stamped on the casing below the reinforcing flange to the left of the differential.

Identifying numbers are:

CG   3.63:1 ratio
CH   3.9:1 ratio
CK   3.08:1 ratio
CL   3.45:1 ratio
CT   3.27:1 ratio

On later cars, the Vehicle Identification Number was stamped on a plate fixed to the right-hand front suspension turret.

PRODUCTION FIGURES

No completely reliable set of production figures for Triumph TR7 and TR8 models exists. Those given below should therefore be treated with appropriate caution. They are for *calendar year*, not model-year.

Year	Market	Figure		Total
1975	Home market TR7	41		
	Export TR7	15,360	Total	**15,401**
1976	Home market TR7	6,923		
	Export TR7	25,820	Total	**32,743**
1977	Home market TR7	8,408		
	Export TR7	14,528	Total	**22,936**
1978	Home market TR7	1,265		
	Export TR7	5,517	Total	**6,782**
1979	Home market TR7	2,451		
	Export TR7	16,207	Total	**18,658**
	Home market TR8	63		
	Export TR8	141	Total	**204**
1980	Home market TR7 fixed-head	1,174		
	Export TR7 fixed-head	88	Total	**1,262**
	Home market TR7 drophead	3,737		
	Export TR7 drophead	6,199	Total	**9,936**
	Home market TR8 drophead	9		
	Export TR8 drophead	2,088	Total	**2,097**
1981	Home market TR7 fixed-head	1,016		
	Export TR7 fixed-head	439	Total	**1,455**
	Home market TR7 drophead	1,781		
	Export TR7 fixed-head	1,421	Total	**3,202**
	Home market TR8 drophead	9		
	Export TR8 drophead	405	Total	**414**
			GRAND TOTAL	**115,090**

Figures from another BL source break down production by factory, grouping TR7 and TR8 together. However, these figures give a production total of only 114,512, or 578 fewer than the figures given above. It is not at present possible to resolve this discrepancy.

Speke built cars (to 1978)	76,081
Canley built cars (1978–1980)	30,375
Solihull built cars (1980–1981)	8,056

All figures are used by kind permission of the British Motor Industry Heritage Trust.

C O L O U R   C H A R T

Note: Triumph paint and trim changes are notoriously difficult to determine and this list may therefore not be exhaustive.

1975–1977 (Speke)

Colours	Upholstery (nylon broadcord)
Brooklands Green	Black
Maple	Beige or Chestnut
Pimento	Black
White	Beige or Black

1977 (Speke)

Colours	Upholstery (plaid cloth with vinyl side panels)
Brooklands Green	Green
Carmine	Red
Flamenco	Red
Inca Yellow	Green
Java	Green
Russet Brown	Red
Tahiti Blue	Red
White	Red or Green

1978 (Speke)

Colours	Upholstery (plaid cloth with vinyl side panels)
Astral Blue (metallic)	Red or Green
Brooklands Green	Green
Carmine	Red
Inca Yellow	Green
Leyland White	Red or Green
Pageant Blue	Red or Green
Russet Brown	Red
Tara (metallic)	Green
Vermilion	Red

Note: These colours were announced for the 1978 model-year. However, so few cars were built as the result of the Speke strike that some colours are rare: the metallics may never have been used on home-market cars.

1978–1980 (Canley)

Colours	Upholstery (plaid cloth with vinyl side panels)
Black	Tan or Navy (Premium edition only)
Brooklands Green	Green (Tan from early 1980)
Carmine	Red (Tan from early 1980)
Inca Yellow	Green (Tan or Navy from early 1980)
Leyland White	Red or Green (Tan or Navy from early 1980)
Pageant Blue	Red or Green (Tan or Navy from early 1980)
Russet Brown	Red (Tan from early 1980)
Vermilion	Red (Tan from early 1980)

1980–1981 (Canley and Solihull)

Colours	Upholstery (plaid cloth with vinyl side panels)
Carnelian Red	Tan or Navy
Meteor Blue	Tan or Navy
Midas (metallic)	Tan
Pendelican	Tan or Navy
Persian Aqua (metallic)	Tan or Navy
Platinum (metallic)	Navy
Poseidon (metallic)	Tan
Triton (metallic)	Tan or Navy

MISCELLANEA

* During 1976 and 1977, BL Motorsport entered a team of TR7s in rally events. The cars ran with 16-valve "Sprint" engines (which anticipated the production of a TR7 Sprint with this engine) and with five-speed gearboxes. The cars were underdeveloped, however, and claimed only one victory during 1976. During 1977, BL poured more money into the rally team, and the TR7s competed in European as well as British rallies. During this season, they ran with disc brakes at the rear in place of the production cars' drums, and were rather more successful. Drivers Tony Pond and Brian Culcheth achieved only one win between them (by Pond, on the Belgian Boucles de Spa event), but were well-placed on several other rallies. However the TR7 rally cars were replaced in the BL Motorsport team during 1978 by TR7 V8s.

* The TR7 V8's career as a works competition car lasted from 1978 until 1980 – three seasons in all. It took a second place on its first outing in the hands of Tony Pond, and won its second event, again with Pond at the wheel. Per Eklund took over from Pond as the principal works driver during 1979, when some experimental work was done with fuel injection. However, major success eluded the team and the 1980 cars reverted to carburettors – four horizontal twin-choke Webers. For 1980, the rally cars' engines put out over 300bhp at 7,500rpm and 245 lb/ft of torque at 5,500rpm.

Tony Pond returned to the BL Motorsport team for 1980 and, despite handling problems, which were developed out over the season, he and Roger Clark were able to demonstrate the car's real capabilities. However, the TR7 V8/TR8 rally programme was halted at the end of 1980. Four-wheel-drive had become permissible in international rallying during 1979; 1980 was the year Audi announced its formidable Quattro coupé; and it was clear that, however good the TR8s might be, they would simply not be able to compete with the new breed of rally car. Tony Pond's TR7 V8 rally car is preserved in the Heritage Collection.

* After the Lynx had been cancelled, BL evaluated a project for 2+2 convertible based on the long-wheelbase Lynx floorpan. This project was known as Broadside. The car would have been extremely attractive: it looked very similar to the TR7 drophead, but would have had the Lynx's styling crease instead of the TR7's controversial sweep along its flanks. Like the Triumph Stag, the car would have had a T-bar for rollover protection. The single prototype, fitted with the 2-litre 0-series engine which was planned for eventual use in the TR7, still survives in the Heritage Collection.

* During 1981, the Sheaffer pen company ran a promotional competition in the UK, in which the prize was a rather special TR7 fixed-head. Painted metallic grey, running on Wolfrace alloy wheels, and kitted out with special decals avertising the Sheaffer name, the car had also been specially equipped by the coachbuilders Wood and Pickett of London. It had been renamed the TRZ, for no immediately apparent reason.

The TRZ had a Lucas LC150 on-board computer with a light-emitting diode readout, and was equipped with a car phone. It had a Blaupunkt stereo system with remote-control keypad on a stalk, and large spherical speakers on the rear parcels shelf. There was also a keyless digital locking system wth the control panel on the rear offside quarter panel, and the interior had been completely retrimmed in velour and Connolly hide. The car was registered as DGH 540 X and still survived in the late 1980s.

* A small promotional run of TR7s was made in conjunction with Coca-Cola and Levi's. The cars were finished in the Coca-Cola colours of red with large white stripes along the sides and across the bonnet, and featured a large Coca-Cola emblem. The interior was trimmed in blue denim, complete with authentic "pockets" in the door panels and sun visors.

* Although the TR7 went ahead as a Triumph, the value of the MG name was never far from the minds of BL's product planners. As early as September 1976, one of the Lynx coupé prototypes was badged as an MG and, during 1980, BL was considering the possibility of badge-engineering the TR7 as an MG to take over from the MGB. At least one styling prototype was produced – a metallic green drophead. Differences from the standard car appear to have been relatively minor: the car wore MG decal badges and had a rather tinny-looking "MG" grille tacked on to its nose. Mercifully, perhaps, the proposal went no further.

* Crayford, the Kent coachbuilding company, converted two TR7 fixed-heads into estate cars in the late 1970s for Page Motors of Epsom.

* International Automotive Design produced a restyled TR7 in 1980 as its first-ever "concept" car. Known as the TRX, the car was converted from an early TR7 fixed-head, registered SFG 666 R.

* Janspeed, the UK turbocharging specialists, initiated a project to turbocharge a TR7 and enter it for the Le Mans 24-hour race. The basis of the car was a redundant BL Motorsport development TR7, and the

partially-developed vehicle was displayed at the 1978 Performance Car Show. However, it did not in fact enter for Le Mans until 1980, by which time ADA Engineering had joined Janspeed in the project. Sadly, the car proved temperamental and failed to qualify for the race, although a timed 201mph on the Mulsanne Straight did demonstrate its potential. A crash during practice for the 1981 Silverstone 6-hour race prevented it reaching Le Mans that year, and a planned entry in the 1982 Le Mans was cancelled because ADA Engineering (who had by then taken the car over) could not drum up enough sponsorship. The car still survives, in private ownership.

* In North America, John Buffum became SCCA Champion in 1977, 1978 and 1979 at the wheel of TR7 and TR7 V8 cars. Buffum also had some drives for the UK-based Leyland Cars team. Bob Tullius successfully campaigned a TR7 V8 coupé in Group 44 events around the turn of the decade.

Leyland's habit of posing pretty girls with the TR7 could become irritating when the car itself was only just visible in the picture! As its registration number suggests, KDU 499 N was a very early TR7 which competed in the Leyland rally team during 1976 and 1977. The car is finished in the Leyland colours of white with red and blue. From 1978, the predominant colour of the rally cars would be red.

The TRZ was the prize in a competition run by the Sheaffer pen company. It had a number of special features, and had been created by coachbuilders Wood and Pickett.

Inside, the TRZ had been retrimmed and had a number of additional features, which were very advanced for the time.

The way things might have been (1): this is a styling mock-up for the MG-badged version of the TR7 which never went into production. The car has the early type of bonnet, without the bulge which arrived when production switched to Canley, and also has the pleated upholstery used on Speke-built TR8s. Perhaps neither feature was intended for the new MG, however: more likely is that the car used for the styling trials was a Speke-built TR8 drophead.

The way things might have been (2): the Broadside convertible had the long-wheelbase Lynx floorpan and an 0-series engine.

The way things might have been (3): when the TR7 family was axed in 1981, there were already plans to revitalise the car. These stylist's drawings show how the controversial curved sweep on the flanks might have been replaced by a neater indentation, and how the roof of the fixed-head car might have been restyled.

PERFORMANCE FIGURES

The figures given here are representative and refer to unmodified cars. They should be viewed as a guide only; actual performance varies from car to car, and there are quite wide variations in performance for the early Speke-built cars.

	Maximum speed	0–60mph (secs)	30–50mph in 4th	Average mpg
TR7 fixed-head, European, 4-speed	112mph	9.6	7.5	28
TR7 fixed-head, US spec, 4-speed	108mph	11.3	–	27.5*
TR7 fixed-head, US spec, 5-speed	110mph	11.2	–	27.5*
TR7 drophead, European, 5-speed	115mph	10.7	8.1	25.3
TR8 drophead, US spec, 5-speed	120mph	8.4	–	15

* US gallons

Note: A collection of contemporary road tests and other articles is published by Brooklands Books under the title, "Triumph TR7 and TR8, 1975–1981".

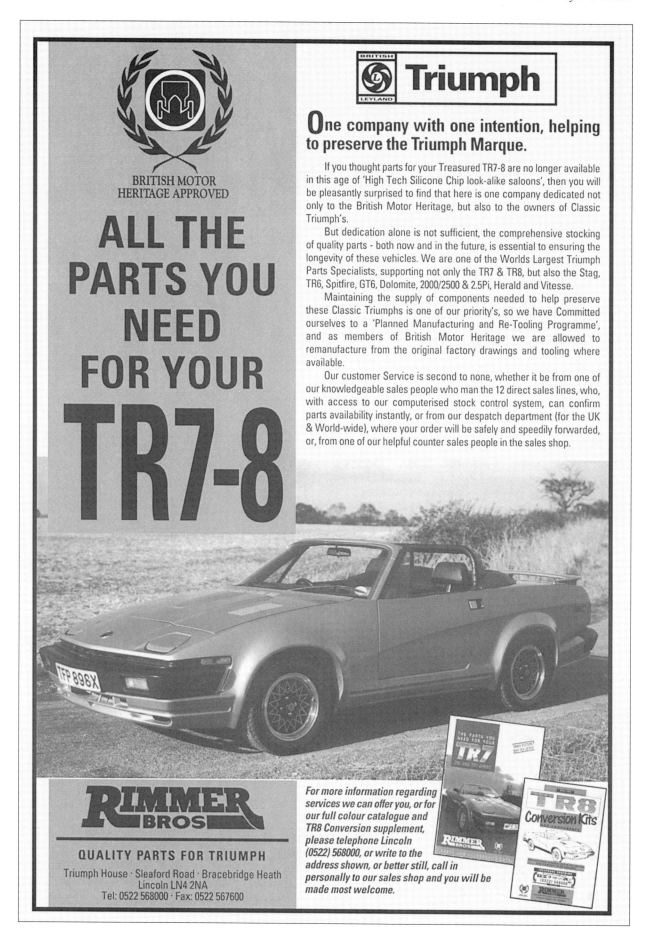